Cosmology According to Biblical and Ancient Models

Flavien Phanzu Mwaka

Cosmology According to Biblical and Ancient Models

The Verses of the Scholars - Volume I

ISBN numérique : 979-10-262-7734-7

EAN papier : 9791026277354

Librinova

Courriel : contact@librinova.com
Internet : www.librinova.com

(FIRST PART : The Pre-Big Bang according to the Bible)

LEGAL DEPOSIT NUMBER /RDC/BNC : MR 3.02003-57107

From the same author in the collection "The Verses of the Scholars".

Volume I : cosmology according to biblical and ancient models. (French version : ed. Exibook) First part : Introduction to Pre-Big-bang (éd. Librinova) Second part : from the big-bang to the living (ed. Librinova)

Volume II : The field of the Universal Cosmological Model (English and French version : ed.Librinova)

Volume III : Pre-Big bang – Big bang - after Big bang (English and French version : ed. Librinova)

Volume IV : The enigmas of the Universe solved in the Universal Cosmological Model (English and French version : ed. Librinova)

Volume V : From Noah to Barack Obama (English and French version : ed. Librinova)

N.B. : online sales and in bookshops

Author's contact : lesversetsdessavants@gmail.com

DEDICATION

This book is dedicated to Moon couple who awakened in me the ideal of perfection. May he bless it and accept it with all their imperfections.

"Man is born pure, it is up to society to make him perfect.".

Flavien PHANZU MWAKA

Epigraphy

"Perfection cannot be reached in ignorance state. One must understand the truth in order to be able to go to the kingdom of God on earth. True Parents have revealed the truth... However, there are still people who live they were many millennia years ago".

Ms. Hak Ja Han Moon : TruePeace N ° 8, January 2015 Pp 6-7.

Acknowledgment

I thank all the promoters of the websites who allowed me to deepen and verify my knowledge to produce this book, without sometimes mentioning the source or the authorization of their authors. Some of these great sites include FuturaS-cience.com, Wikipedia, Encarta, Trustmyscience.com and many more. We also thank all those who are surprised to find their names in this book through their various statements in documentaries, conferences and certain articles on the net. We apologize to them while reassuring that some of their comments were only quoted as testimony and not as arguments. For, everything that is written here is the product of the heart, for God only communicates with us through our heart, which signifies our spirit according to the Holy Scriptures.

I would also like to thank Mr Augustin MABWONO MUKWAMA Anguishi of the School of Translation and Interpretation of the National Pedagogical University of Kinshasa/DRC for his proof reading in English despite all his other activities. However, the translation from French to English has caused the text to lose some of its content. This is due to the translators' unfamiliarity with scientific language and even religion. Thus, any future translation must refer to the original text, i.e. the French version.

Finally, I would like to thank The Bible societies which these do not hold against me making the verses read directly to my readers rather than paraphrazing them or referring them to a book. In science, it's like stopping the engine and stepping out to open the hood. We used the Louis Segond version and the King James version.

Flavien PHANZU MWAKA

PEAMBULE

In the French version, the first and second parts are one book. But afterwards I considered it too voluminous for a book on cosmology which calls for a lot of effort of understanding and self-sacrifice on the part of the reader. Thus, in the English version, and in other versions that will follow, the first part will only include the introduction to the before big-bang.

In this first part we prepare scientists to understand biblical language and non-scientists to understand scientific language. This is what we are going to do up to chapter V. The point concerning the before Big-bang, will only be dealt with from chapter VI onwards.

Until then, researchers have speculated about the before Big-bang. But, having discovered it in the first cosmological book, i.e. the Bible, and in the writings of the ancient philosophers, with all the great ensembles that characterised this period, I thus become the first godfather of this discovery.

From now on, all the theories that will be built on the pre-Big-bang and that will not take into account one of these biblical and ancient components will only be a slip. This will be confirmed after the development of this volume in the other volumes.

This gave birth to the Universal Cosmological Model which is the scientific restitution of the scheme of creation as contained in the book of Biblical Genesis. We will come back to it later.

THE AUTHOR

PREFACE

Good research works with undeniable religious, scientific and cultural depth. The field covered is tough and defensible in the light of reliable material evidence This text should be published as a trial balloon for readers unfamiliar with cosmological material combined with biblical evidence.

I would like the publication to be partial (3 or 4 volumes) in successive editions and not all at once because of the rigidity of the addressed material.

Once again thank you and good luck!

Dean of the Faculty of Arts

National Pedagogical University (UPN)/RDC.

Foreword

God the Almighty is first of all the Creator. from the invisible, he brought out the visible one. This Volume attempts to recreate that event scientifically. This recreation, we call it "The Universal Cosmological Model", which is a new revolution within the scientific development. Therefore, this book is neither a confessional book, nor a doctrinal one, but it is the book of cosmology. Even though we could conclude that God is the Creator of Cosmos, it is now a part of cosmology and no longer exclusively of religion, because this book provides convincing arguments on that.

There is one fact that everyone should know : considering that we all are the creatures coming from one common origin, namely God, the universal truth is inscribed in each person's mind. The more we think about a topic subject, the more mankind search for its truthiness in his consciousness, and this appears as a raw material in his everyday life. That is why there is always similar ideas between two authors who did not even know each other before. In such cases, the only difference that exists among researchers is in terms of the enunciation, especially between an atheist and a believer researcher. Then, it is the most enlightened one, who will one day place the findings in time and space, as it is currently the case with the Universal Cosmological Model contained in this book.

In the Antiquity, cosmological thoughts and theories were the work produced by the philosophers and religious.

These were the ideas expressed by the greatest thinkers concerning the origin and evolution of the universe. Religiously speaking, their authors attributed them to divine revelations. At that time, they thought that the universe was limited to only a few horizons. But, guided by their intuitions and scientific reason in the one hand, and by God's wisdom on the other hand, they came to conclusions which are being confirmed by modern scientists today. The introduction of mathematics and the experimental instruments moved this discipline more and more towards the exact sciences and converted it into theoretical physics.

But, as it is written in the Bible : "my words will never pass away" (Mt 24:35), the contribution of this book entitled "cosmology according to biblical and ancient models" is to resuscitate the earliest and the most structured stories, although they have been rejected by the scientists, namely the biblical scheme about the world creation. The resurrection under discussion here is supported by both the ancient patterns also buried by modern history - namely the Ancient Eastern Cosmological Model in the one side, and the Ancient Western Cosmological Model on the other one.

Our present study is based on the biblical scheme of creation which has never been taken into account by astrophysists, that is, by the scientists. Through this diagram, we are going to restore it scientifically under the label of the "Universal Cosmological Model". It gobbles the "Cosmological Standard Model" which considers the Big-bang as a singularity, so as not to enter God's domain which we call "the field of Moses" or before Big-bang according to science. From now on, this book gives birth to a new branch called : Astrotheology.

God entrusted the Universe to the man by saying, "Fill

the earth (universe) and subdue it" (Ge 1:28). Therefore, He cannot oppose the scientists approach in their quest about the origins of the Universe. Questioning about the universe's origin is also incorporated part in our genes. So, inspire of the human limited capacities - which are still in the early stages - we could not escape far from the biblical history of the creation. Our task is to restore it originally in scientific language according to the Universal Cosmological Model. Thus, this book now makes all cosmological models that do not include the before Big-bang and the Big-bang itself as obsolete one.

The creation schemes of the antiquity come to us from a time when we believe people were ignorant without knowing any physics laws. This so- called ignorance is simply due to the difference in terminologies used traditionally comparing to those used by the contemporaneous scientists today. According to Moses, his scheme was rejected simply because his truths are recorded in a so-called sacred book that is, based on faith rather than reason. This book is called the Bible.

The verses inventoried in the collection : "The Verses of the Scholars" are not written for evangelization purposes. They are rather selected here as purely scientific quotations. They do not deserve to be read in a meditation way like we used to do while praying through the words of God, because they are situations and physicochemical states to be subjected to the scientific experimentation. So, they have nothing to do with our religious faith. It is the same reality for a certain number of the so-called sacred books cited here. All of them are surely scrutinized for scientific and not doctrinal purposes.

This is how, through this work, biblical verses, the words of scholars and their writings, as well as the writings of religious will often come back during our analysis. As we have already said, all these excerpts found in this book are used for scientific reasons, but not for religious reasons. Insofar as human knowledge and experimental instruments are limited, those of the future will probably not be, because religion being the domain of the mind, i.e. of thought, it has always been ahead of science.

This book often takes into account only the Bible verses related to science. Some of the concepts come from the book "The Presentation of the Divine Principle" by Reverend Dr. SUN MYUNG MOON.

In the book "The Divine Principle", Reverend Dr. SUN MYUNG MOON has clarified all the parables that have caused scientists to reject religion.

This book is the first non-doctrinal book from a student who has received this teaching, but has nothing to do with the doctrine of the church of Reverend Dr. SUN MYUNG MOON.

It is an individual work, the fruit of my research and analysis, and this, through my little scientific knowledge in relation to Moses, and my direct relationship with God.

Therefore, all the comments and remarks contained in this book are binding on me. It does not deal with any church doctrine, and I write only for my own pleasure as an autodidact writer and researcher, like a musician who is ready to glorify God through a Gospel song. I do not have any justification to render to neither the Church nor to anyone. It is just like an inventor who sharing what he has discovered to the world of science.

My motivation was only to fulfil my childhood dreams in presenting proofs through science and observations that is God who is the true Creator of the Universe. But, I am far from having achieved my goal, because there is still a big gap between the spiritual world (heaven) and the formation of the shapeless and empty earth or the universe.

(Ge 1 : 1-2).

This book; "The Verses of the Scholars - Volume I", is a book designed for the public. It scientifically restitutes how the universe came out, its evolution based on the biblical scheme of the creation. This restitution dealing with the evolution of the universe according to the biblical diagram of creation, we have called it "the Universal Cosmological Model". We said evolved instead of created, because what goes beyond the Uncertainty Principle (energy-particle relationship) is difficult to demonstrate through experimentation, that is, the shift from spiritual world (as explained in the Principle of Creation, branch of the Divine Principle book by Rev. Dr. Sun Myung Moon) into energy in a scientific way.

I do without the classical system of the "Foot notes method" to reserve only the "Author-date method in the text" system. Since the majority of extracts and arguments are quoted from articles, websites and scientific documentaries. For the sake of modesty, I'm going to use most of the time the polite expressions "we" and "our", instead of "I" and "my". These expressions have nothing to do with teamwork. By using "I" or "me", I feel very haughty, while it is with all humility that I submit this work to humanity.

The originality of "The Verses of the Scholars" collection is the revelation that God works much more with scientists and researchers, than with religious men. If pastors and priests are there to bring peace to the hearts of individuals, the happiness that people expect to experience in the Kingdom of God on earth is in found through in the responsibility of scholars.

For example, during the Corona virus pandemic lockdown that brought mourning to all mankind throughout the world, all believers' prayers were asking

God to inspire the specialist of the medical care to find an efficient vaccine which must stop immediately the spreading of the virus incurable up to now. Being believer or atheist's searchers, God has no other choice than to work with both of them, because it is through them that the third (last) promise of God's benediction will be fulfilled, namely "control the world" (Ge 1:26). We wish to reveal to our readers how the world which represents the Universe was created according to the "Universal Cosmological Model", and we will prove it.

The Universal Cosmological Model which is explained in this volume is a masterpiece and an original idea. It does not deal with any extension of a doctrine. Also it is not based on the inspiration from an existing Scientific Model. It does not add anything to the Universe either, like all cosmological models. The universe has its Creator who is God and not any models.

Cosmological models attempt to explain the creation either literally or mathematically and they constitute analytical schemes leading to its formation. Hence, the laws of physics are mathematical equations that we used to explain these models. Thus, we define Biblical Models not as the interpretations, but as scientific analysis for the biblical scheme pattern on the creation of the universe. This analysis has given birth to the Universal Cosmological Model.

The French astrophysicist Aurelien Barrau wrote :

"No doubt that there is something different about the elaboration a scientific model. One of the descriptions of the world is obviously the creation of the cosmos. To set up a theory (or even making sense of an observation) requires great liberty and ability for invention ".

(Aurélien Barrau : Big-bang et au delà, balade en cosmologie, Dunod, Paris).

An African proverb says : "*Once you see a snake, you no longer ask questions anymore, you kill it with the piece of wood you hold in your hand"*.

Thus, the realization of this model in the environment where there is neither no one to bring your assistance, nor libraries to consult books relating to cosmology, the existence of the faculty of astrophysics is a lure, we were obliged, a part from the biblical knowledge, to run again on the websites to get some useful information from certain articles and documentaries which provided me with some elementary theoretical knowledge on the fields mentioned above.

However, without the book about the Divine Principle written by Rev. Dr. Sun Myung Moon (Presentation of the divine principle : ed. Culture et Paix, Paris, 2013), in which he explains in a literal way, sometimes in a scientific language, the world of the cause, this book could never be published. The documentaries and lectures findings on websites have helped us much to complete this book and allowed me to gather supportive comments.

Thus, this book is the first literature that used as references articles from websites, lectures and documentaries. If the scientists accept this as a reliable source of inspiration, this is a pleasure for me. Indeed, the researchers explain explicitly much better during the presentations of researches in front of the audience than in the books, because here, there are a lot of physical and economic constraints. In addition, the scientific thought requires demonstrations rather than a book, and this is the reason that there are many scientific journals.

The collection "The Verses of the Scholars" appears to bridge the gap between religious books in the one hand, and scientific ones, on the other hand. These religious books we are talking about here are the

Bible, the Koran, the Divine Principle of Rev. Dr. S.M. Moon, the book of "Transformation" which talks about Eastern cosmology and the articles on the thoughts of a lesser-known Philosopher named Empedocles, whose writings on cosmology are said to be the foundation of the Ancient Western Model.

Astrophysicists have no other choice than to master the explanations contained in these books, in order to solve some enigmatic, and to lay a mathematical base. Thanks to knowledge acquired in the explanations of these books, that we succeed to fulfil a gap on the before Big-bang, then this allows us volume et accidentally within the world of astrophysicists by bringing them back the Universal Cosmological Model so that they could scrutinize it.

All of the names mentioned in chapters one and two of the book of Genesis in the Bible have deeper meanings in astronomy than in religion. The religious men are bound to what they read, without being able to probe in a deep analysis, simply because they are already struck by the limitation spirit" Do not add anything, and do not remove anything away". However, these sayings in the Old Testament concerned the laws that Moses dictated during the Exodus to Israelites people (Deut 12:32), and in the New Testament also, in the book of Apocalypse or the Revelation of John (Rev 22:18). These recommendations led Jesus to the Cross, and were the cause of the execution of several philosophers as well as scholars by the Church because of that. Let's read this :

Deut 12 : 32 'You shall observe and do all the things that I have commanded you; you won't add anything to it, and you won't remove anything "

Rev 22:18 "I declare it to everyone who hears the words of the prophecy of this book (Revelation) : If anyone adds anything to it, God will add to his punishment the plagues described in this book; 19 And if any one takes anything away from the words of the book of this prophecy (Revelation), God will cut off from him his part of the fruit of the tree of life and of the holy city, described in this book ".

In the collection "THE VERSETS OF THE SCHOLARS", we are going to point out the scientific equivalences of the names which were quoted literarily in the biblical scheme of the creation, thus forming the universal Cosmological Model in cosmology. This enlightens the ingredients necessary for astrophysicists in order to tell us about the reliable results of their experimentation.

In addition, this work should persuade people to consider Moses as the « scholar of all scholars". All research on the origins of the Universe leads to God. This is indicated by a lot of questions that always come up in people's mind.

Therefore, I invite my readers not to be troubled by the Bible verses cited in this book. Indeed, they are used here as scientific citations, and will help to increase the intelligence of the reader who will know how to transpose them from a literary language into a scientific one. We have made efforts to not go neither deeply into religion matters, nor to make this book as a religious manual.

In the collection of books "The Verses of the Scholars," By highlighting religion, we discover the superiority of Jewish wisdom in the antiquity and even up to these days.

The author.

General Introduction

Why do cosmology according to biblical and ancient models? first of all, a model is a diagram, a theoretical representation aiming to report a process, the existing relationships between various elements of a system. (Cfr:vinsvignesvignerons.com/Geologie/Une - breve - histoire - de - la - planete - terre).

For scientists, the title of this Volume could be "The Cosmogony according to Biblical and Ancient Models". As we said it so far, cosmology seeks to apprehend the Universe from a scientific point of view as the whole of the material distributed across the space-time, whereas, cosmogony aims to establish the theory of the creation in terms of philosophical or religious basis.

The scientific restitution of the biblical scheme of creation within the Universal Cosmological Model, moves it from cosmogony to cosmology.

In all books dealing with the history of the Universe, we do not find anywhere these two universal wisdoms; namely, the biblical scheme of creation and the formation of the universe according to the ancient wisdom. However, in the light of my research, it seems clear that these models revealed by God to their authors are more true than one may think, to the point that the contemporary science is lagging behind in discovering certain things that have contributed to the formation of the universe according to ancient thoughts.

Indeed, this work has just enriched the scientific world with the contribution of two models hither to ignored in cosmology, namely, the Universal Cosmological and the antique cosmological models. This sends researchers back to investigations. We agree that the outcomes of the scholars will lead to the recognition of the theories which are in this book and, therefore, to the scientific admission of the "Universal Cosmological Model»

The authenticity of the Universal Cosmological Model compared to other scientific models in cosmology lies in the fact that it deals with the before Big-bang, passing through the Big-bang until the appearance of man on earth, and this in a more harmonious and coherent way than the current science. Therefore, the Universal Cosmological Model swallows up all the partial cosmological models that scientists have offered us so far. "Swallowed" means that it captures them and puts each one at its right place in order to allow the scientists to provide us with the coherent diagram.

For example, annihilation, the uncertainty principle of quantum mechanics and the String Model are engulfed in the Universal Cosmological Model and should be classified in the before Big-bang.

Today, science uses limitative theories to explain the universe. They are restrictive in the sense that they set an absolute limit on human knowledge in different fields. So, to explain the period after the Big Bang, science uses the theory of general relativity. To go down up to the Big Bang, it uses quantum mechanics which contradicts certain values of the theory of general relativity. To explain the Big Bang, scientists use string theory, which is not yet harmonized. And to explain the appearance of the elements from the energy contained in the space, they use the Principle of uncertainty.

To agree with this model, I said, has nothing to do with neither the spirituality, nor philosophy or metaphysics, because here we are in front of the physical process that led to the formation of the visible universe.

If, for theologians or the common run of people, this volume is a set of sentences, but for physicists and astrophysicists, everything that is written here is part of the physical phenomena which occur inside the stars or in the space, that is to say, in the universe. And it is done thanks to the natural forces coming from God; which only astrophysicists know. So this is part of physical phenomena, and some of them are verifiable in atomic physics laboratories.

For science, before Big Bang, all reasoning was only the speculation. Nevertheless, there is a direct witness who is and who was there before the Big Bang and who tells us what existed before; that is God. The lack of a proper interpretation of the biblical scheme of creation has left cosmology in deep doubt. Today, thanks to this book, the discovery of the interpretation of how the events before the Big Bang took place opens a new door for scientists' research.

Chapter I :
Erudite

In order to understand a work and know how to criticize it, it is better to know first its author. Thus, the story about the biblical creation from which we formulate the Universal cosmological model is attributed to Moses. It is in his biblical writings that we find the great secret of the Universe. Thus, it is important for us to know man according to science, who with the publication of this book, we are having his name immortalized in the universal pantheon, and God recognized as a living reality.

Another great scholar without whom no one would have known the meaning of great parabolas in the Bible is Rev. Dr. Sun Myung Moon. Without him even this book, which allowed me to unravel the mystery of how the universe functions, this book could not be written. And, finally, it is also with humility that I want to express my motivation through this chapter.

Section 1 : The Prophet Moses

1. The prophet MOSES : The greatest scholar

– (1250 to 1030 BC according to historical sources)

– (1593 to 1473 BC according to research made by Jehovah's Witnesses Biblical Society)

According to historical sources, as we have discovered them in encyclopedias, Moses, who is often referred to as a myth, lived from 1250 to 1030 BC. However, according to Jehovah's Witnesses Biblical society, whose research is more reliable concerning the history of religion, Moses would have lived from 1593 to 1473 BC. (Cfr : the book " The Great Babylon has fallen apart").

These years which mark the period of slavery in Egypt are effectively included in the providential periods of the history of the Restoration, as developed by Rev. Dr. Sun Myung Moon in the Divine Principle book.

(Cfr : Presentation of the divine principle : ed. Culture et Paix, Paris, 2013, Pp 382-392).

Therefore, when we wish to restore to today the diagram of the biblical creation written by Moses, is to decipher the truth hidden in the first book of cosmology written 1500 BC. on the papyrus. Where did this great understanding of Moses come from?

Ex 33:11 *"The Lord used to speak to Moses face to face, as a man speaks to his friend".*

If there is yet any other man to whom humanity will have to pay homage to forever as it is the case to Jesus Christ, to Muhammad (let peace and mercy be upon him), to Buddha and to Confucius, it is well and truly Moses. But, did he exist or not, a myth or a reality?

The three great monotheistic religions claim to be originated from Moses. They called him Mosheh in Judaism, Moses in Christianity and Mûssa (Moussa) in Islam, he occupies a primary place in the hierarchy of characters in the Old Testament : he is the unifier of the Jewish people, the one who has received the Ten Commandments (or Decalogue) from Yahweh and transmitted them to his people.

Born in Goshen, in the Ancient Egypt (15th century BC), Moses belongs to the tribe of Levi, one of the twelve Hebrew tribes that migrated to Egypt in the 17th century BC.; son of Amran and Jochebed, his brother, Aaron and sister Miriam.

Few days before Moses was born, reported the scriptures, the Pharaoh of Egypt ordered the persecution of the Hebrews by killing all their newborn males. To save her son, Moses' mother places him in a basket which she places in the reeds on a bank of the Nile (Ex 2 : 4). The child is picked up by the pharaoh's daughter, who has pity for him, then adopts him, and raises him like a prince. She gives him the name of Moses which means according to the Bible : "saved from the waters".

(Cfr : "Moïse." Microsoft® Encarta® 2009 [DVD]. Microsoft Corporation, 2008).

Moses is therefore, according to Jewish tradition, the founder of their religion - Judaism, which is sometimes called, Maism, that is, the religion of Moses. He was the first man to be named the "man of God" in the Bible, and the only one man to be named so in the Torah. He is also mentioned in the Qur'an under the name of Moussa as the Prophet and Messenger of Allah.

For Jewish and Christian monotheism, Moses is both the author and the main writer of the Pentateuch books. He is attributed having, under the divine inspiration, written five books of the Bible : Genesis, Exodus, Leviticus, the Book of Numbers and Deuteronomy in which he has told us about his own life and his ancestral patriarchs' life, as well as the people of Israel. He is one of the prophets whose thought influences without discrimination all mankind around the world, especially in the domain of natural law, sources of positive rights, canon law and ethics, in the Muslim world as well as in the Christian one.

He was the prophet and guide who led the children of Israel out of Egypt, the land in which they were subject to slavery, and it is his powerful hand that the "divine wonders" (the ten plagues of Egypt and many others) would have been carried out in order to allow the liberation of the Israelite people and their departure to the promise land of Canaan.

History is transmitted to us by those who do research or those who lived through these events. But, who lived the events from before Big Bang until the formation of man, so that they could relate it in a great detail to Moses as we will see in this book ? Concerning the before Big-bang, for example, even the scientist's calculation deduced that this sphere is insurmountable ; and that no law, neither physical nor mathematical ones, can give them answers.

Moses named God as his source of inspiration ; and I line up behind him without forcing that conviction on anyone. But at least whether you're an atheist or not, you agree that one plus one are two. It means that the truth is always the truth for everyone, regardless of the obedience of the person speaking it ; whether you are an atheist or a religious person. The theories of the Belgian priest, Abbot Georges LEMAITRE, were initially criticized because he was religious. These theories have become up to these days the foundation of the Standard Model in cosmology, and supported by the majority of astrophysicists, being atheists or believers.

Other prowess's of Moses will be developed in the other Volumes. For example :

– In science, Moses is the first to domesticate the solar energy, the first to use chemical and other bombs, during the struggle against his enemies.

– In medicine, he dictated preventive rules known as impurity, for those who disobey God's instructions. He warned the Israelite people against the practice of cloning, and homosexuality. Knowing that the most difficult diseases to cure are those contaminated by blood, he warned against its consumption and the use of all practices that can promote it. Note that consumption is not exclusively by mouth.

– In botany he warned against hybridization and dictated the characteristics of plants that cannot be eaten and when a plant can be eaten, too. He dictated where we can find plants that have medicinal molecules that can cure incurable diseases.

– In culinary art, he dictated the characteristics of animals that can be eaten and those that people should be abstained from because its microbes, bacteria and viruses.

All of these prowess's are developed in the volume V with supporting Bible verses. Following case of disobeying these instructions it would result in dreadful illnesses, including CORONAVIRUS which is plaguing mankind at the time of publishing this book.

Thus, the day when people are going to discover the vestige that this man has left and approved that, Scientists will give God his right position again he deserves. On that day, we won't say the following sentence : "nature has provided". Henceforth we will rather say "God has provided". This is also the aim of the books in the collection "The Verses of the Scholars".

Moses is indeed a living personality and the time has come for mankind to reflect on what he presented for the happiness of mankind, this is what we have developed in Volume V. It only remains to recognize him as the scholar the "Scientist of Scientists", a title that scientists use instead of prophet, a name used by the ancient religions.

Egypt at that time was a developing nation in the spiritual and technological fields. The remains attest to this. Pharaoh was considered not only as a physical authority, but also as a spiritual one. Therefore, the princes should benefit from this education and initiation too.

MOISE, whose Pharaoh's daughter took care of him as her own child, also benefited from a princely education. The Pharaoh's daughter took a Hebrew woman, who was none other than Moses' own mother, as a foster mother for him. She instilled in him the Hebrew tradition and religion. His quadruple education, namely initiation into the religion of the Hebrews, the religion of the Egyptians, science and the army, made him above his contemporaries in wisdom and intelligence.

He also benefited from the art of war, since the Pharaoh should also be the chief of the army. This justified that after he had been living in Egypt, the Hebrews conquered all their enemies during their passage to Canaan. So, Moses was not only a prophet, a scholar, but also a warrior; all these title was not exploited by the Torah and the Bible so as not to tarnish his quality as a prophet. The following verse attests it in these terms :

Ac 7:22 *"Moses was taught all the wisdom of the Egyptians, and he became a great man in words and in deeds."*

Thus, from the first to the third volume, we are going to discover what Moses has bequeathed to us in his various writings or recommendations, and what neither religious nor scientists have understood or have been able to exploit in depth up to these days.

As far as the Genesis of the Universe as narrated in the Bible is concerned, there has not been development so far. Even the theologians do not dare to publish it, and pastors refrain from giving sermons based on the first chapter of Genesis. It's something that makes people doze off, even in church.

Indeed, it is a great scientific subject, which requires a lot of thoughts and calculations. But human as a son or a daughter of God will succeed, provided that he/she accepts humility. Regarding humility, Moses submitted by completely effacing himself as a searcher and scholar, but he put God at the center of his life, as the only subject of his discoveries. So he often returns to this by repeating the expression "God says". so, his statements are recognized as revelations. Scientists, who do not like the word revelation, have put his discoveries aside, and, ipso facto, man himself has been forgotten.

Moses knew that he had brought nothing to the universe with regard to cosmology or the science that studies the creation of the world according to religion, and the Universe for scientists. Thus, all that Moses had as a revelation was discovered in his laboratory thanks to his thoughts during the 40 years of his life at the palace of the pharaoh and the 40 years of his exile in Midian, he has dedicated them directly to God. That is why there is the use of the famous phrase "God says" throughout the book of Genesis. This means that the situation was under divine control. This humility, by attributing his research and results to God, earned him an elevation from God Himself.

Ex 4:16 *"Aaron will be your spokesman and speak to the people for you.*

Then you will be like God, telling him what to say".

In others words, in the eyes of the Hebrews, Moses was not considered only as a prophet, but, he was also like god for them. We understand the reason of the crucifixion of Jesus Christ because some of his words were in contradiction with those recommendations dictated by God through Moses.

Until now, no scholars or great politicians have publicly dedicated their glory to God, apart from Presidents John Fitzgerald Kennedy and Barrack OBAMA in the United States of America who used to refer to the name of God whenever they speak about their great achievements in the domain of research about the Universe or the space exploration.

Moses, in describing the period of cosmic deflation as "one day", his words were criticized as a myth, and his schema of the creation of the world as legendary one. The Rev. Dr. Sun Myung Moon is the first researcher since the creation of the world to bring out the book about the usage mode of the cosmos. He called it "The Principle of Creation". Unfortunately, he was criticized as a brain washer. The American physicist Allan Guth offered the theory of inflation (theory according to which the first elements of the Universe, namely hydrogen, were created in the space of 10^{-43} seconds to 3 minutes), he was called a scholar because of his university degrees' diplomas. If he had not got one, in spite of all his experimental demonstrations and formulas, he would have been treated as crazy. Anyway, to be called as a crazy man is the feeling that drives any astrophysicist before publishing an idea or a new theory.

Finally, there is the proof that testifies the existence of God, and it can only be found by doing one's own introspection, that is, by analyzing oneself as it is written in the pestle to the Romans :

Rom 1:20 "Indeed, the invisible perfections of God, his eternal power and his divinity, are seen as with the eye, since the creation of the world, when one considers them in the things that has made. They are therefore inexcusable". And the mankind is his most perfect creature

Section 2 : The Rev. Dr. SUN MYUNG MOON (1920-2012)

Throughout this book, we are going to talk about the Principle of creation. It is one of the branches of the book entitled "The Divine Principle" which constitutes the teachings of Rev. Dr. Sun Myung Moon taught within the former "Unification Church", which has now become "Holy community for Celestial Parent". The Principle of Creation is the fundamental teaching that allows us to understand well the origin and function of the universe. This is the base of the Universal Cosmological Model.

2.1 His life

Born in 1920 in Korea, the Rev. Dr. Sun Myung Moon graduated from high school in electrical engineering department. He was at the age of 16, then as a fervent believer in the Christian Presbyterian Church, Jesus appeared to him telling him to complete his mission. He did not understand first what this means, because somewhere it is written "everything is accomplished" (Jn 19:30). Rev. Dr. Sun Myung Moon began searching deeply for that truth, but he only understood the purpose of this message through Jesus' first words, namely "The time has come, and the kingdom of God is near" (Mk 1 : 15), and these last words which ended by "today you will be in paradise with me" (Lk 23:43).

None of the messengers of God after Jesus ever have understood the difference between these two spheres : the kingdom of God and paradise. This has never attracted the attention of theologians so much, as they are limited by the physical horizon. The Rev. Dr. Sun Myung Moon was the first to be able to point out the difference in his teaching found in his book entitled "The Divine Principle".

When he was 19-year-old, he went to Japan to study engineering at Waseda University. But the real purpose of his coming to Japan, according to his bibliography, was not to study technology. Above all, his experience in Japan allowed him to broaden his vision of the world, to deepen his understanding about the problems encountered by people across all countries, and to compare them with the Principles that he had discovered together with all the scientific achievements of the modern times. His studies were interrupted because of the war.

Despite this, his experience and discoveries allowed him to explain the origins of divine energy; cause of the function of the universe scientifically. His outcome constitutes a new revolution in theology and philosophy, and, why not in science too. His teaching brings together both disciplines, namely religion and science, which were separated many years ago. And today we have just put a link between both sciences, through the collection of books called "The Verses of the Scholars".

For having dared to evangelize in a communist country, the Rev. Dr. SUN MYUNG MOON was sentenced in the North Korea during his youth (28 years) to 5 years of forced labor in a camp in Hung Nam, where he had to fill and carry bags containing ammonia powders and toxic fertilizers. It was an ordeal reserved for those who deserved capital punishment, but who should be killed later on. The maximum survival time for the strongest men was one year, but the Rev. Dr. Sun Myung Moon exceeded it, and for that, he obtained the certificate of the most enduring man.

While the survivors were being executed one by one, he

escaped from the prison after two years and eight months, on October 14, 1950, thanks to a bombardment of the United Nations forces under the leadership of Marc Artur who broke the walls of the prison. During his life time, he was grateful to this general, because he was going to be executed the next day.

While he was still with all his vigor and intelligence, at 92 years old, his lungs could no longer play their role of filtering oxygen because of what he endured in the Hung Nam camp. It was at that age that he physically left us.

The Moon couple : Rev. Dr. Sun Myung Moon and Ms. Hak JaHan Moon.

In this Volume, we focus more on the contribution of Rev. Dr. Sun Myung Moon in the field of science as well as on that of Moses.

Section 3 : Flavien PHANZU MWAKA

3.1. This work is not a revelation

If you want the truth about something to be never considered by science, just call it "revelation". Scientists like the use of the words "discoveries" or "research". Thus, the creation story as narrated in the Bible - and it constitutes the only absolute true, despite the parabolas unveiled in this book –it has never been the subject of scientific research, because it was considered as a revelation.

As an illustration, the first book of Genesis consists of the prophecies about the creation of the universe. Moses has seen things globally, but thanks to his intelligence he had succeeded to classify them in order. In Genesis II, he again put his intelligence into contribution by focusing only on what happened on the planet earth. And from there, we see that the scientific scheme is respected as we restore it in this book.

The other example comes from me. God blamed me and ordered me to tell everyone that "the only true scheme of the universe is the scheme of creation according to the Biblical model, and people should be inspired by it." Scientists want solid evidence or arguments in order to believe it, that is say, they need the details on that. This is what I have tried to do in this book, by introducing myself into a field that was not part of my training. But, as being believers, we know that the source of intelligence is God. Once one centers it towards Him, He is going to open his door and the inspirations come.

Anyone who touches the origins of the Universe touches the plums of God. That is why God is interested in me among the 7.7 billion people in the world in the year 2020. This is also why astrophysicists limit their research to analyzing the universe from the Big Bang only, i.e. they stop where the laws of physics cannot be applied so as not to enter God's domain. Anyone wishing to study the pre-Big Bang period must either follow what the Bible says on the first day of biblical creation.

A quotation from Bishop Saint Augustine extracted in the scientific documentary "Cosmos - Beyond the Big Bang", written by Matt HICKEY, says this : "If one finds an interpretation of the scriptures that seems to oppose what is established (science), this interpretation must be reviewed".

It is the same approach that I proceed through the collection of books "The Verses of the Scholars" in which some would risk to call me a heretical. Being heretic or not, this judgment has nothing to do with my relationship with God. Galileo said : "To declare something heretic by the fact of believing in what is proved as the truth can be considered undoubtedly harmful for souls". (Cfr : J. Baker : 50 keys to understanding astronomy, ed. Dunod, p11).

A hundred years ago, the big bang belonged to the greatest discovery in cosmology in the 20th century. The fact that I discover this reality in the Bible, that is to say, in the writings ancient times, it is considered as the greatest discovery in the 21st century, because it will rectify a lot of data in cosmology. Up to now then, even believers, scientists or not, were skeptical about the Big Bang theory. Scientists supported this theory in order to reject God as the creator of the universe. Thus, the fact that the Big Bang is biblical confirms that God is the true Creator of the universe. It revives the debates that the Belgian scientist Abbe G. Lemaitre stifled by forbidding Pope Pius XII to promote his theory of the primitive atom.

Our promotion at L'OBP / France : on my right, Mr. Marc Breton; the director of the Observatory. (September 2019)

In fact, in 1951, Pope Pie XII affirmed in a speech during the Pontifical Academy of Sciences audience that "the true science (...), the more it progresses, the more it discovers God". He even specifies that "science has succeeded in being itself the witness of this 'fiat lux' acronym.

"The reference to the new cosmology extracted from Einstein's equations and Lemaitre's outcome about the origin of the universe, is clear".

Now, according to Lemaitre, as we said at the beginning, such a "concordance" between the revealed and scientific truths is prejudicial for science even for religion. According to Lemaitre, we cannot at anyway make a connection between both fields because they are parallels, that is to say, they are the doctrines of the "two realities", the supernatural reality taught in the Holy Scriptures in the one hand, and the natural one discovered by the scientific reason on the other hand. After the reaction of Lemaitre, one year later, during a new intervention, Pope Pie XII has adopted the point of view about the separation of both domains. (Cfr : At the origin of the Big Bang : Abbe Georges Lemaitre (Humanity.fr : Science and religion).

Now, it is almost the time came so that science and religion come together again as one. For example, it is not science which has revealed that the Universe was initially empty and formless, but it is religion (Ge 1 : 2 "the earth was formless and empty"). It is not science which has discovered that the planet earth was suspended in an empty, but it is religion (Jb 26 : 7). This book of Job probably dated since the 5th century B.C., after the return of Israelite people from exile to Babylon. Science only confirms that the earth is round, and it is suspended in an empty after Fernando Magellan's world tour in 1519, finished by one of his lieutenants named Elcano. (Cfr : "geographic exploration." Microsoft® Encarta® 2009 [DVD]. Microsoft Corporation, 2008).

Let us read the following biblical sentence :

Jb 26 : 7 *"God stretches out the northern sky, and He hungs the earth in empty space".*

Ec 1 : 7 *"Every river flows into the sea, but the sea is not yet full. The water returns to where the rivers began, and starts all over again. "*

The north on the void, meaning that the confines of the universe are only space, and everything material is suspended in this void, revolving around the different orbits. Word said and written in the 5th century BC. The other verse, written by King Solomon several centuries B.C., stated that the earth was round even before science confirmed it.

This paragraph constitutes only an overview concerning the great biblical discoveries that are revealed in the collection books : "The Verses of the Scholars" collection.

The "Universal Cosmological Model" represents what is set out in this book, while "the Biblical Models" represent the set of models of other scholars who will support the Bible message, even if my explanations differ from theirs. The Universal Cosmological Model is, therefore, the translation into a scientific language about the schema of creation, as written in the first and second chapters of the Genesis book. This scientific translation will hence forth be the foundation of all Biblical Models.

Also, while writing this book, I discovered several natural and physical laws by myself, and most of them are scientific laws that I did not know before. I prefer to keep on with my conclusions and explanations in order to be understood by my readers. Indeed, the astrophysicists' jargon is too complicated to be used in a literary work. What someone discovers himself, he can explain it well to others more than what you are taught brothers. Richard Feynman confirms that by saying, "What I can't create I don't understand".

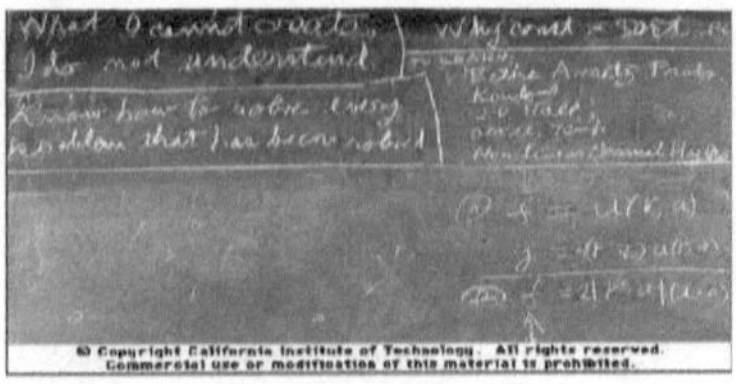

https://qph.fs.quoracdn.net/main-qimg-87833c78a604ff07

Explaining Feynman's position, one internet user wrote this in Quora.com, which I very much appreciated it : "*Feynman meant here that understanding something is not just about working through advanced mathematics. You also have to have a fairly intuitive notion to explain it to an audience that cannot follow the detailed derivation*".

(Cfr : Mark Eichenlaub in Quora : What did Richard Feynman mean when he said, "What I cannot create, I do not understand") ?

Richard Feynman approach is indeed what I do.

Thus, as I said it so far, I bear full responsibility for any ideas, thoughts, and even errors if there are any, contained in this work. I wrote it without a help of from any supervisor or any teachers. Like Moses, I also lacked true scientific words or expressions to explain what was really going on in my ideas.

My aim in this book, without having any experience in literature, is to expose only the Universal Cosmological Model in its broad lines by restoring scientifically, but in a literal way, the biblical scheme of creation.

(Working session at the Observatory of Baronnies Provencal in France / September 2019).

After receiving God's injunction, I had to manage to convince scholars that the history of biblical creation is the only true cosmological model.

As I have said it, and I will keep on repeating it, the Universal Cosmological Model does not concern God or religion. It is the scientific restitution of the first Cosmological Model found in the papyrus, and whose author attributed all the effects to divine intervention, it means that, the need of God's will. Whether one is a believer or not, its development has nothing to do with faith.

Thus, this book as a scientific-religious literature as one can category it constitutes in cosmology, a new scientific revolution after 100 years of the reign of the Big-bang Model. The contents will initiate other major discoveries by giving a new direction to astrophysicists and cosmologists. It is going to do the same for theology and philosophy. The ideas and the revolutionary explanations of the biblical verses contained in this book will be supported by the theses of the students pursuing their studies in theoretical physics, in theology and in philosophy, and will allow a new advanced progress in these sciences.

In the world of titles, compared with regards to the books "The Verses of the Scholars", I am only a self-taught researcher in cosmology, but basically, that everyone draws his conclusion.

Chapter II :
The keys

These are the keys that blocked religion within the science. Indeed, by reading the various scientific and cosmological writings published so far, the universe like an egg hatched by the force of energies during a phenomenon called the Big Bang. For classical cosmology, before the publication of this book, it is said that all started there. According to the Universal Cosmological Model, which is the scientific restitution of the biblical creation scheme developed in this Volume, the Big-bang was only the rebound of a Universe in compression, but whose expansion spread the resulting material.

At this point, we want to demonstrate theoretically that the Biblical Model should be part of cosmology, by restoring the first and second chapters of Genesis book. This restitution has given birth to the Universal Cosmological Model.

Indeed, one cannot deny or hide a truth simply because science rejects it or scientists ignore it. So, let's read the truth about the beginning of the universe as it is written in the bible :

Ge 1 : 2 *The earth was formless and empty; there was*

darkness over the face of the deep, and the Spirit (a wind) of God was moving over the waters.

Ge 1 : 7 *And God made the firmament, and divided the waters which were under the firmament from the waters which were above the firmament. And so it was.*

Ge 1.9 *From the waters below earth appeared*

Ge 1.14 *From the waters above the lights were created.*

Section 1 : Branches for studying the universe

1.1. Astrophysics and cosmology

Nowadays, the term "cosmology" refers to the branch of astrophysics which is concerned with the origin, nature, structure and evolution of the universe taken as a whole. Astrophysics is defined, according to Joachim Herrmann, as the study of the universe by physical means. (Cfr : J. Herrmann, Atlas of Astronomy, ed. Stock, Paris, 1976).

We add that astrophysics is the literal expression of the mathematics of the Universe. Astrophysics has taken the ideas of certain philosophers in the development of its history, while rejecting those philosophers whose thoughts have been recorded in the sacred books, as in religion, this is the case with the diagram of creation contained in the Bible, the Koran and other religious books.

To this day, to explain the mysteries of the universe, scientists rely on what they see, and draw inferences from it. Me, on the other hand, I relied on what we do not see, that is to say the confrontation of the energetic forces at the beginning of the universe, to arrive at what we see, qualified by scientists of observable universes. Their setbacks and my advancement made us land in the same airport called Big-bang. From now on, we will all have to un-paste together to arrive at the real knowledge about the Universe, by uniting the points of view of astrotheology and astrophysics.

1.2. Theology and astrotheology

Indeed, theology and science have always ignored that the link between God and man is the universe. Theology studies both the aspects of God and man while forgetting that before this first one integrates into the second, God had to first create the mankind from the dust of the universe, called "star dust" by scientists in the one hand, because they have analyzed the situation uphill. Moses, on the other hand, called man "dust of the earth" (Gen 2:7), having based himself downstream. With the Universal Cosmological Model, the biblical creation story now becomes in astrotheology "the biblical scheme of creation".

By adopting the Universal Cosmological Model as a part of cosmology, each concept should be understood, not literally, but according to its comprehension in cosmological science - rather than the theological one. It is because, in theological science, each concept is interpreted according to its etymological meaning either in Greek or Hebrew. According to the Universal Cosmological Model developed in the collection of books "The VERSETS OF SCHOLARS", we probe, not the word, nor the concept used, but the main idea of the author.

In Astrotheology, the main and inescapable branch is the Principle of Creation; a branch of the Divine Principle of Rev. Dr. Sun Myung Moon. The scholars of these two branches, theology and cosmology, always feel the desire to give sermons. Those of scientists are called popularization. Thus, in astrotheology, cosmology is the detailed and scientific study of the biblical scheme of creation. It must be said that up to the present day cosmology has not yet integrated the pre-Big-bang, that is, what happened on the first biblical day. This integration, I make it in this book "The Verses of the Scholars, Volume I".

1.3 Metaphysics and astrotheology

How does astrotheology different from metaphysics? Metaphysics (Metaphysics = beyond physics) is a philosophical science that first questions the existence of things or events as they appear to us, and then tries to describe and explain what really exists (Cfr; Wikipedia; Metaphysics). Thus, man is at the center of his answer and God may or may not be evoked.

In astrotheology, on the other hand, God is the foundation of all things and everything in the universe functions according to these dual characteristics (Rom 1:20). It integrates the study of cosmology as one of the external aspects of God.

The Big Bang itself is very far from God's domain. To knock off the door to the Big Bang was supposed to enter into metaphysics. However, the creation history in the Bible has nothing to do with metaphysics. Moses is telling us in detail what really happened, but in a language of that time which has been distorted today from its original texts little by little by the translators. Perhaps the true interpretation of the diagram of the first and second chapters of Genesis will come from the astrotheologians. Especially with the publication of this book, the study of the scheme of creation according to the Bible is going to become even more interesting.

So, as stated so far in the title of the book, cosmology for our study includes God as the Creator of all. The name of God appears here as being the opposite or the opposite of "the natural accident". It is as if there are two names; namely, "God" for us as believers, and "natural accident" for scientists. However, if thanks to "God", we can explain the cosmological phenomena that science can experience them, making hypotheses one day which could be translated into theories, laws and formulas, then our approach remains scientific.

In the Science Documentary : "HAS THE UNIVERSE EVER BEGINNED ?"a KLEIN LECTURE on Science & You, the scientist is wary of the singularity of the Big-bang which he considers as a transitory stage, and demonstrates that the study of the before Big-bang cannot be considered as metaphysical, but physical.

Section 2 : Discrepancy between science and religion

2.1. Biblical language

The scheme of the first and second chapters of Genesis in the Bible is a true story in cosmology. History, because it is told in such a way as to please the reader instead of using the complicated scientific language of 1,500 BC. The Physicist Paul Dirac said : "*In science researchers try to tell people something that no one has ever heard of it before with terms that everyone can understand easily. But in poetry, it is exactly the opposite.* "(Cfr : J. Baker : 50 keys to understanding astronomy, ed. Dunod, p.69).

In my opinion, I think it's quite the opposite. It is in poetry and literature that we find comprehensible terms accessible to all, rather than in science. For example, instead of talking about galaxies with an audience of non-physicists, I will better have understood if I use other terms like 'other worlds', and finally conclude in the end that it is about what we call "galaxies".

All we are seeing as theories in present day cosmology is just the scientific development of what the Bible has revealed. The Theories of the Bible have been called a revelation and set apart, because to this day mankind has never discovered a large telescope or a laboratory that Moses used. And, in addition, Moses' discoveries are recorded in a sacred book. But, fortunately, until then, science only keeps on confirming the biblical revelations as we are going to see it in this book.

If the writings of Moses were recorded in another book other than a religious one, the mankind would have admitted already him as being a scholar. This is what people should do now by scrutinizing the cosmological scheme of Genesis, and it is not the supplication, because we demonstrate it through this Volume.

Indeed, most of the religious writings are made of verbal speeches and sometimes expressed spontaneously. But, once they are written and published, they become a literary and scientific work ready to be analyzed. Which means they ready to be criticized or to be investigated. It is like any religious song or poem, produced in a medium, and made available to the public, becomes a work of art. So their authors are called artists.

There have been theories many years ago, and we still have other scientific theories tomorrow that will contradict the biblical revelations regarding the creation of the Universe. But, all of these kinds of theories have been built up in the gap, and they will always be challenged by other scientists and researchers without knowing it. This is the case, for example, with the theory of evolution, which nowadays stumbles upon the missing links, and with which Louis Pasteur has confused his contemporaries.

After the failure in his attempt to create bacteria from sterile broth, Louis Pasteur has, in fact, demonstrated that the living beings, even the microscopic ones, are not born from nothing, but they are only developed 'from germs which already existed in the atmosphere, it is just as physicists have stated that there is a vacuum which contains energy and therefore virtual particles.

Thus, in the verse "The earth was formless and empty" (Ge 1 : 2), there was already energy in this void which is considered as God's essence as such (Cfr : Presentation of the Divine Principle, ed. Culture. and peace, Paris 2013, Pp 29-30). This is what makes the difference between the spiritual world called heaven and the physical one that we call the earth (universe). Hence the Biblical verse "In the beginning God created the heavens and the earth" (Ge 1 : 1).

2.1.1. **An ancient and parabolic language**

There is one fact that we need to know : the biblical language used by Moses was spoken in such a way to convey the message to his contemporaries. The vocabulary meanings were not as rich as it is today. In addition, the divine revelations or dictations are inner phenomena captured by the soul. That's why we sometimes have to take appropriate expressions in order to make them understood to the audience. This means that Moses had not mastered that somehow, and God began to dictate to him the message which he should spread.

For example, in a seminar organized by the MOON Couple Church on Unification Thought in 1988, I could no longer fall asleep when the speaker brought up the point on the creation of the universe. He touched on the point that has always fascinated me, namely, how the material appeared from the intangible, knowing that God also works in a scientific way, which is summed up by the famous phrase of Moses : "God said".

God himself had to intervene in my dream. He has showed me how the transition of the elements is created or takes place on the basis of electric discharge caused by violent thunder lightning and generating strong energies. In my dream, I saw the thunder and its lightning deeply in the Universe in an area of significant electrical activity, followed or preceded by ice. At that time, I was completely ignorant of the structure of the sky that there are glacial planets, and it is very cold. I didn't even know that there exist some electrical activities which occur there. All what I see in the sky with my naked eyes here in black Africa in which there are no telescopes, are the stars, the moon and the sun. We are going to come back to the interpretation of this dream when we will talk about the formation of dust in Volume II.

How can we share such an idea in a world full of scientists in physics and astronomy? Even the believers themselves may consider me as a heretic person. But nowadays, what is my relief, my idea when I read on the Futura science site about the experience of scientists who almost demonstrated what I saw in the dream. Since I do not belong to their field, I prefer to report exactly the following extract :

"We are not proposing a scenario for the appearance of life," as one of the researchers named Marco Saitta has explained in Futura-Sciences. We suggest a new avenue to specialists in prebiotic chemistry. "] (Cfr : 09/10/2014 at 11:38 by Jean-Luc Goudet, Futura-Science).

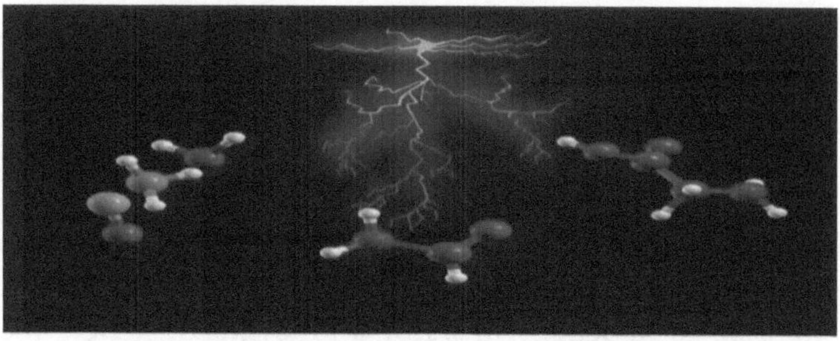

"Artist's impression of the effect of a flash or an intense electric field on three simple molecules, on the left : water, ammonia and carbon monoxide (from top to bottom). The reaction produces formamide first (COH-NH2, center) and then the simplest amino acid, glycine (right)". © A. Marco Saitta (IMPMC / UPMC) and Franz Saija (IPCF / CNR)

Also with Miller's experiment, other scientists who learned the conditions in interstellar space, realized that amino acids, called molecules of life, were produced simultaneously in space by electric discharges. They were transported to earth by comets (blocks made up of dust and ice wandering in space). It still fits to my dream of lightning on the one hand such as thunders produced and ice (comet) on the other. However, the real explanation will be unveiled in the following chapters, as proved by science.

In addition to these experiments which confirmed what God showed me in my dream, I discovered again to my great surprise that two scientists have already developed the concept of electrical universes by demonstrating the natural force of electrical phenomena in the processes of creation. This paragraph is developed in the point dealing with the Universal Cosmological Model versus the Electric Universe in Volume II.

To return to our subject, we would like to recall people that the spiritual language, contained in books determined as sacred ones, such as the Bible, the Koran, the teachings of Buddha, Confucius, the Mayan writings or other esoteric books relating to spirituality, are very different from literary or everyday language. The same word can have different meanings.

For example, when the Bible speaks about death, it is about the separation of man with God through sin, but it is not a physical death. Thus, in the following sentence "the wages of sin are death" (Rom 6:23), we see the saints dying young and the bigger sinners die around 99 years old. And when the Bible talks about eating, it is the context of the sentence that determines whether it is something taken by mouth or intercourse. Thus, in the verse "This is how an unfaithful woman acts : she eats and wipes her mouth, and says:" I haven't done anything wrong" (Pr. 30:20). We understand directly that it is about sexual intercourse and that we do not have to resort to the meaning of this verb in Hebrew or Greek as the theologians do.

When Jesus was physically on earth, he said "I have come to set the captives free" (Luke 4 : 19-21). His sentence, taken literally, would mean that he should go to prisons to free people. However, this was not the case. Even John the Baptist in his captivity doubted him, asking if it was he or someone else should be expected (Mt 11 : 3).

Today all Christians understand that these sentences have spiritual significations. That is to say, some biblical sentences, although pronounced in everyday speech, have a different meaning, compared to the true reality. It is a spiritual language called a parabola, and whose book "The Divine Principle of Rev. Dr. Sun Myung Moon reveals the meanings.

2.2. **The concept of time and labor in the Bible**

The history of the Bible begins with Genesis in which the first chapter speaks about the creation of the Universe.

This story, which should be used as the foundation for cosmology, in the contrary was rather neglected by scientists, thinking to find discoveries that contradict it. However, in the light of laws and discoveries associated with this science, the omnipresence of God is more and more asserted as being the Creator of the Universe. Prior to the publication of this book, scientists believed their findings were increasingly alienating God as Creator, and the religious were at a loss.

Moses, who is credited as the writer of the Genesis story, did not have a laboratory or astronomical glasses, not even telescopes like our scholars nowadays. And, yet, everything he wrote is being confirmed up today. Chronologically speaking, the order of the creation of things in the biblical scheme and the appearance of things for science is almost in harmony. When we say almost, that doesn't mean that science is right, especially since to this day we continue to find fossils that set back or change history.

But why do scientists stray from the Bible's creation story? There are two main reasons for this question, but which are misunderstood by them, namely, the non-existence of the time notion and labor in the biblical schema of creation. But, we are going to show that both notions do indeed exist in the biblical story of creation.

2.2.1. **The existence of the notion of the time in the Bible**

The story of Genesis shows us that there exists one God who has created all things instantly, like a magician; that is wrong according to our literal analysis. There is, indeed, the notion of time in the history of creation and this notion is well defined in the same Bible.

Many religious believe that bringing the notion of time into the biblical schema of creation is a way of diminishing the omnipotence of God ; that is not true. In everything that God has done, there has always been the notion of time. The omnipotence of God is revealed through the perfection of things that he has done, regardless of the length of time that it took, it took place in one day, within the space from the evening till the morning : "*there came one evening, and then came one morning* "(Ge 1 : 5-27).

2.2.1.1. **The meaning of "one day" in the bible**

2.2.1.1.1. Meaning of the concept "Day" according to the Bible

Despite the biblical verse : "*But there is one thing, beloved, which you must not ignore, and that is that before the Lord a day is like a thousand years, and a thousand years are like a day*" (2 P. 3 : 8), no Christian theologian has been able to relate this verse to the days in the pattern of biblical creation, believing that the latter was the work of a word. It took another revelation from God for the Moon couple for enlightenment to emerge based on this verse.

Ignorance of the Bible's explanation of the days of creation meant that no theologian could stand before scholars with convincing arguments. Now that the MOON couple has lifted the veil, we can stand before them and speak the same language.

Indeed, thanks to the book "The Divine Principle" written by Rev. Dr. Sun Myung Moon that I have had the explanation about the days of creation, otherwise this manual could not be written.

Throughout the history of biblical creation, there is one phrase that keeps coming back, namely "there was evening and there was morning" to mark both the beginning and the end of a creation, or a day according to the Bible. And that is the end of an era, according to science. Certainly, scientists will ask the question of what this symbolism represents. The end of a day according to the Bible is equivalent to an era according to science; which can last for millions or even billions of years.

Indeed, considering the books contained in the Bible as a whole by Christians, whose author, God, is like a hero in the shadows, the biblical day as quoted in the first chapter of Genesis is well defined in the same Bible in two verses :

Ps 90 : 4 "*A thousand years to God are like one day, they are like yesterday already gone, like a short hour in the night.*"

2 Pet 3 : 8 " *But there is one thing, beloved, that you must not ignore, and that before the Lord a day is like a thousand years, and a thousand years are like a day.* "

These supporting verses are always cited as a justification in postponing Christ's return, when he said it two thousand years ago that he is coming back soon.

Rev. Dr. Sun Myung Moon is the first religious man to have associated them with the biblical scheme of creation in the Creation Principle; a branch of the book of the Divine Principle.

There are two very important terms in these verses, namely "to ignore" and "like". First of all, we should not ignore the notion of time. Secondly, there is the comparative term "like" : "a day for God is like a thousand years". Here, the word "equal" has not been used, but rather "like". This simply means that a biblical day can represent as many days, as many years as a cycle or an era can take.

So to say that creation was made in six biblical days has nothing to do with our twenty-four hour days. Rather, it represents six great stages which, as science has discovered, took thousands, millions, even billions of years.

In addition, the sequence of events as narrated in the first chapter of Genesis is the same with what science has discovered so far : the mankind being the latest creature to appear on the earth.

A striking illustration which shows that these biblical days are not like the ordinary ones, but they included periods, so, it is during the three first days of creation found in the first chapter of Genesis that there was not yet creation (appearance) of luminaries to mark the notion of time. We notice that the luminaries' appearance took place in Ge 1:14-19 during the fourth period of creation.

Gen*1:14 " God said, "Let there be lights in the expanse of heaven to divide the day from the night; let them be for signs to mark times and days and years;..."*

19. " Evening and morning came : it was the fourth day."

So, in terms of what criteria did they fix the first three days, since the notion of time began on the fourth day? It was fixed according to the outcome of the evolutions that were unfolding in the Universe.

In order to be able to talk about the notion of time, three ingredients should be taken into account, namely, a star spinning around another and an intelligent person who is able understand to that. The rotation of the star on itself and its translation or revolution around another, thus create the quantifiable notion of time. Now, as we said it above that during the three first days of creation, there was no existence of the stars. So the days of which we speak in the Bible in chapters one and two of Genesis, constitute periods and not, our ordinary days of twenty-four hours.

This unquantifiable time of the universe is not recognized or not taken into account by science. It's called the Pre-Big-bang period. Science only takes into account the events that followed after the Big Bang. In the Bible, these post-Big Bang events began just at the time of the "separation of water from the waters" (Ge 1 : 6). Finding the precise location of the scientific Big Bang in the Bible is one of my greatest discoveries. And this is a great revelation for scientists and religious people. The Big Bang was revealed in writings dated back to 1,500 BC. We will come back to this in the following chapters.

So, before the Big Bang or *"before the separation the water under it from the water above it"* according to the Bible, the notion of time was very complicated. Science does explicit about it, because it has put a cross concerning this period. But in the Universal Cosmological Model that breaks this cross, we have pointed out three steps : an empty universe full of darkness (Ge 1 : 1), then came the manifestation of the power of God over the waters (Ge 1 : 2), known under the scientific name as the action of energies upon the original materials. By the first Universal Force, (super gravitation according to science), this power of God will lead to the third step, namely, an extreme warming generates rays. This is what is biblically specified by the following sentence : " *God said : Let there be light. And the light was.* (Ge 1 : 3).

That is how all that dark period from the before Big Bang until the appearance of the first light according to the Bible is called the first day (Ge 1 : 2-4). The Big-bang itself occupied the second biblical period (Ge 1 : 6-8) and the period which deals with the star formation occupied the third and fourth biblical days (Ge 1 : 9 and 14), without however having reference stars to calculate these times.

This development of thoughts agrees with the thought of scientists that we may read :

"*Cosmologists would like to specify the dates of these phenomena, and of course study them, too. But how do they get some information about what has happened in the dark ages when there was no star shine as a reference, and how do they observe these very first stars that are called "Population III" stars ?* (Cfr : futurasciences.com. First stars in the universe : signals finally detected, Laurent SACCO.

03/02/2018).

All of this is to confirm that there is a very fine notion of time in the history of biblical creation, and the days we are talking about have nothing to do with our twenty-four hour days Thus, the notion of biblical time, emerged in the Principle of Creation, branch of the book of the Divine Principle of Rev. Dr. Sun Myung Moon, supported by my arguments in this book, shows that there are no discrepancies between science and the Bible.

2.2.1.1.2. **Evening and morning according to the Principle of creation**

Ge *1 : 4 "... Evening passed and morning came ..."*

Why did the Bible use both concepts "evening and morning" instead of referring to other times' expressions used during the day ? Once again, we resort to the Principle of Creation, a branch of the Divine Principle book of Rev. Dr. Sun Myung Moon as follows :

"The concept" day "in the first chapter of Genesis does not refer to the normal twenty-four-hour of a day, as for the expressions" morning "and" evening "they cannot be interpreted according to the ordinary meaning any more. Consider this passage in Ge1 : 5 "... evening passed and morning came : the first day." After a step of God's creation deeds which had lasted one evening and one night, the coming morning or harvest time was considered by God as belonging to the first day time. Still in Ge 1, God commanded : "there was evening and there was morning : the second day ...the third day ... the fourth day ... the fifth day ... the sixth day. (Divine principle, level 4, P.31)

– The *"evening"* means, according to the divine revelation contained in the divine Principle of Rev. Dr. Sun Myung Moon to Which I share the same ideas ", the period when God created or ceased to create. The period during which God ceased to create something or an object until the period when the ideal of creation was reached : it is a period of growth. It is represented by *night*, the moment between the evening and morning. Scientifically speaking, this period is known as "evolution." Something under goes the process of evolution when it has not yet reached the final shape or maturity required. So the "morning" represents the period during which the ideal of creation was realized, that is, the period of maturity or achievement.

To understand all this, let's consider the three stages of the creation or formation of the mankind. When God said : let create human beings like our image ; here we are in front of God's will. Then came the time when it is written "God molded man" until the time when man became a human being, this represents the evening. After that, there was an evolutionary process which gives birth to an intelligent human being ready to receive the command of God, and to decide on what his God wants. This is the growing season or the night ; what science calls the evolutionary period.

– *The morning* represents the fulfillment of the ideal of creation or the time of maturity, which allows God to go ahead to another stage. Hence the biblical sentence "Evening passed and morning came ..." (Ge 1 : 4), to mark the beginning and the end of a period.

So we have :

God commanded (said) : which represents his will

Evening : palpable embryonic formation of his will, the seeding moment or the origin.

Night : growth of the created thing or evolutionary period.

Morning : the accomplishment moment according to God's final ideal or the harvest time.

You should know that each of the three steps cited above can take a thousandth of a second (inflationary time of the second day) to several thousand, millions or billions of years until the achievement of the ideal that God seeks, and it is represented by a biblical day (2Peter 3 : 8). This has been explained previously as we can still read them :

2 Peter*3 : 8 But there is one thing, beloved, that you must not ignore, and that is that before the Lord a day is like a thousand years, and a thousand years are like a day.*

Let's examine an excerpt from a speech showing that the notion of scientific time is also taken into account by religious in relation to the six biblical days which are only periods and that this is not only an interpretation of Rev. Dr. Sun Myung Moon.

This is the speech given at the Pontifical Academy (1951), and then at the astronomers (1952) by Pope Pius XII for encouraging Abbot Georges Lemaitre who searched and stated to have put forward the idea of the "primitive atom" which is going to become the foundation of the Big-bang theory :

"On the other hand, the avid truth spirit rightly insists by asking how material has never been able to come to such similar state, so inconceivable for our experience today, and on seeking what has preceded it... With the same limpid and critical view that the facts are examined and judged, he foresees and recognizes the work of the almighty creator omnipotent inside whose virtue aroused by the powerful Fiat pronounced billions of years ago by the Creative Spirit, was deployed in the universe, calling into existence, in a gesture of generous love, the material overflowing with energy. It seems, in fact, that the science of today, going back millions of centuries, would succeeded to become the witness of this initial Fiat lux, at that moment during which nothing emerges, with the material, an ocean of light and radiations,

whereas the particles of the chemical elements separated and assembled into millions of galaxies by themselves. "

It is not important to remind you that the Belgian Abbot Georges Lemaitre asked Pope not to interfere into science matters, while reminding him that his discovery was purely a scientific one, and it had nothing in common to do with religion. But, till today, I congratulate Pope Pius XII for having seen some divine inspiration within this discovery, because the notion of "primitive atom" stated by Belgian Abbot Georges Lemaitre is well and truly biblical one. It is his discovery from the Bible that allowed me to write this book with divine encouragement.

2.2.2. The existence of the notion of "Labour" in the Bible

There exists well and truly the notion of labour in the history of creation and this notion is well defined in the same Bible by the verbs to do and to form. Many religions think that introducing the notion of labour into the history of creation is another way of reducing the omnipotence of God. It is not true. In everything God has created, there has always been the notion of time and work. For example, when we read the verses below :

Ge *1 : 3-5 : "Let there be light! And light appeared... So evening passed and morning came : that was the first day".*

How will a mighty God able to dictate the appearance of things with his magic word wait a whole Bible day to proceed to the next step ? It is unthinkable. With our false understanding about God, things could have been done in one instant rather than during six biblical days, which it is not the same case here.

There is always a work from the beginning of the thought until the completion of the act. We can take the easiest example. You leave in the morning to go to the work, your wife tells you that she will prepare the food you like. The idea and the pattern of food is already in his mind. You return from service and find the table ready as promised. Your wife has fulfilled her promise as she said it in the morning, and you are happy to eat the food you love in the evening (thus, evening passed and morning came).

For man, it is as if the woman had said : Let there be food. With such a thought, one cannot appreciate the great work undertaken by the woman, just as with the simultaneous appearance of things, one cannot appreciate the great labor of God.

In the previous paragraph, there is the thought and the final step. But, between thought and the final stage, there is and was work. To get to this final stage, a lot things had happened. And even this food, while it is boiling, it releases a certain gas into the atmosphere that can cause stomach aches if they are ingested by humans. So, it is the same with the sentence phrase "God said", which represented a thought, and the things appeared. But between the two actions : "be and appeared", there was a great work that lasted thousands, millions and even billions of years. The most striking illustration of the notion of time and labour, and which justifies the effort made by God, we see that in the formation of man and woman. (Ge 2:7-8,19).

Gen*2 : 7-8 "The LORD God formed man from the dust of the earth, breathed into his nostrils a breath of life, and man became a living being. 8Then the LORD God planted a garden in Eden on the east side and put the man he had formed in it."*

Gen*2:19 "And the LORD God formed every beast of the field and every fowl of the air out of the earth, and brought them to man to see what he would call them, that every living creature might be called what man would call it."*.

Here, we are in the presence of a God who is the Creator of all, and who did not give himself a rest time.

Gen 2:21 *"Then the LORD God caused a deep sleep to fall upon the man, and he fell asleep; and he took one of his ribs, and closed the flesh in its place.*

This being so, we see that in everything God has done in the biblical scheme of creation, there has been the intervention of time and work. The misunderstandings between religious and scientists concerning these two notions are thus cleared up. Otherwise, how else can we justify this warrior's rest that God gave himself on the seventh day?

Gen*2 : 1 ""Thus were the heavens and the earth finished, and all their host. 2 God finished his work which he had done* *on the seventh day, and rested on the seventh day from all his work which he had done. 3 God blessed the seventh day, and sanctified it, because on that day he rested from all his work which he had created and done. "*

There is a very important word here that the Bible insists on it to point out the concept of labour; it is the verb to do. He used to repeat it himself insistently three times to erase the image of an effortlessly fulfilling magical word of God. And in Ge 3 : 3, the verbs "to create" and "to do" are mixed together, to show that this creation is the result of an effort. A reasoning which until then escaped all theologians' minds. Let us read :

Gen*3 : 3 God blessed the seventh day, and sanctified it, because on that day he rested from all his work which he had created and done. "*

Section 3 : The Voice of God

3.1 God says

We want to come back again to that magical sentence of "God commands" and "the things appear" in the biblical creation diagram. To whom was God speaking or commanding ?

When we scientifically analyze the creation process today, scientists have discovered that there were no longer any cuts from the first particles appeared or were created, until the formation of man (Ge 1 : 2-27) - although science sometimes speaks of spontaneous symmetry breaking at a certain moment, or some accidental mutations. This is why scientists do not believe in the evolutionary predestination of creation.

There has been a harmonious cyclical evolution in the order of the appearance of things. For example, everything that breathes could not appear without the presence of water.

With that, we should ask ourselves the question to know whether God really dictated the appearance of things. Through the light of the analysis that is done in this book and what science has been able to discover, we see that this was not true. Because, if that was true, we would have a DNA structure different from plants and animals ; that is not the case. For this reason, we can say that things have developed together, and nothing has emerged independently out of the others. From the creation of the first atom, up to the creation of humans, everything is linked. And the man who is the last creature appears as the microcosm of the cosmos having within him all that is physical, all energy and all spirituality in the cosmos.

The word "evolution" can be a pejorative concept to religious people because it was previously used by those who wanted to challenge the Bible. But what makes the difference between our ways of thinking and the fellowers of evolution? This comes effectively from the sentence of "God says" which is used in Genesis, demonstrating the predestination of the creation scheme.

Indeed, the partisans of evolution exclude God as Creator in their though, while we accept Him (Ge 1 : 1). And we say that the first phenomena that resulted in the original particles before we get into the Big Bang, which you will understand in the following chapters, would occur thanks to the ideal of God. This is what justifies Moses' use of the sentence "God said".

Let us consider a concrete case of evolution, but under the will of God :

In Gen 1 : 3 "God said : *Let there be light! And there was light*»; now, this light appeared, not because God our Father said there should be light, but rather it was a consequence of the spirit of God (energy) on the waters (original materials) in the abyss (kind of black hole) (Gen 1 : 2). So, it was a phenomenon obeying physical laws. But, since they were part of the creation scheme as predestinated by God, Moses spoke of "God said" before their appearance.

So why did Moses use the expression "God said" every time, while we were dealing with evolutionary and physicochemical phenomena? This meant that the formation of the universe was done under the divine control, and that was part of the ideal of God. So the sentence "God says" has nothing to do with a magical sentence, and that is what has hidden the biblical scheme of creation in cosmology at that time. As we have illustrated it before, it is like a woman who is cooking and she does not need to repeat all the time that "the water boiled, the food became pasty" ; all the schemes are already inside her mind. But, to shift from one step of evolution to another in the Bible, Moses used the sentence "God said" to show that it was not "a simple natural coincidence of hazard" that was happening.

The Universe, which science considers to be a vast laboratory, is part of God's cauldron. Moses called this pot an abyss, which is for us the original black hole. It is by there that all started until its explosion or expansion in a Big-bang and which was completed not by the formation of the man, but by the achievement of the perfection by Jesus Christ to whom somebody returned this testimony : " Here is the Man " (Jn 19 : 5).

Scientists will tell me that evolution ended with the appearance of Homo sapiens. This is a debate that I myself am opening up, because no scientist has ever affirmed it in a book, nor has any religious person ever noted it. It is the Rev. Dr. SUN MYUNG MOON who explained the three great steps qualified as the three great blessings in the book of the Divine Principle ; stages which constitute the phases that a man must naturally go across, but with the sharing of responsibility.

Scientists are not wrong insofar as their analysis is limited to the physical point of view whereas for us religious people, God's outcome for man is when his body and mind become or will become united, centered on God, making man as a perfect creature (Mt 5:48). And Jesus was the first man to reach this sphere (Heb 5 : 7-8). Read :

Heb 5 : 7-9 "*In his life on earth Jesus made his prayers and requests with loud cries and tears to God, who could save him from death. Because he was humble and devoted, God heard him. Even though he was God's son, he learns through his sufferings to be obedient. When he was made perfect, he became the source of eternal salvation for all those who obey him".*

If Moses used the sentence "God says", it is to say that it was in God's ideal principle to achieve the desired results as science finds them. Thus, "God says" is part of the inner will of God, while the action initiated is a part of the voice of God.

Ge *1 : 3 God said, Let there be light!* (Inner will, ideal);

Ge 1 *: 3... And there was light.* (Action or voice) 4 *God saw that the light was good.*

Ge 1:26 *Then God said : Let us make man in our image (inner, ideal will);*

Ge 1:27 *God make human being in his own image* (Action or Voice) 1Ge 31 *God looked at everything he had made, and he was very pleased.*

Here we are talking about the expression "God says," which is used in the biblical scheme of creation, not in the whole Bible. God has spoken to many people directly in the Bible; and outside the Bible; including me. His way of addressing people is like an audible word. But when we look around, we notice that the person to whom God is addressing is the only one to hear it himself. This means that the voice of God is interior. And, the concerned person, for whom it is intended, hears it through his soul, by his spiritual senses, sometimes believing to have heard it thanks to his ears.

Let us take two examples among many others : it was during the baptism of Jesus by John the Baptist. The latter was the only one to have heard the voice of God, and the phenomena that took place (Mt 3 : 16-17).

The something happened with Samuel who heard the voice of the Lord, thinking that it was the prophet Elijah who called him (1 Sam 3:8-9). We read :

Mt *3 : 16-17 "As soon as Jesus was baptized, he came out of the water. Then heaven was opened to him, and he saw the Spirit of God coming down like a dove, and alighting on him. Then a voice said from heaven : "This is my own dear Son, with whom I am pleased ".*

1 Sa*3 : 8-9 "The Lord called Samuel a third time. He got up, went to Eli, and said, "you called me, Here I am." Then Eli realized that it was the Lord who was calling the boy, so he said to him, "Go back to bed; if he calls again, say, Speak, Lord, your servant is listening. And Samuel went back to sleep".*

These examples are given to believers because scholars already know that to attribute the audible word to God is to attribute physiological functions to him.

By His will and power, God had to work for things to appear and be formed. This is confirmed by the biblical sentence that mentions the end of every creation as "this is good. Such an expression can only be said when one contemplates the successful work that one has just done and not what one makes appear in a magical or miraculous way.

3.2. God said, God made, God created and the spontaneous symmetry breaking

By symmetry, we mean the conservation of properties, a kind of overall stability even within a movement. The breaking of symmetry therefore supposes that a system in which something that was being preserved suddenly ceases to do so and gives rise to an unexpected phenomenon.

The history of creation for science is the story of spontaneous breakdowns of symmetries. Moses, in the Bible, divided them into six days, which are part of the six periods of creation. But, within each period, we find unexplained phenomena. For example, if we take the first day, or the period minus one in science (i.e. not yet recognized), we might ask the following question :

In the verse "God created the heavens and the earth (universe)"; how did the physical (the earth) appear from the spirit? According to science, there has been the breaking of symmetry somewhere. Symmetry was a stable sky that was preserved. And suddenly, part of the sphere of the heavens filled with the spiritual particles, unexplainably were transformed into a particle and physical antiparticle in "a formless and empty earth."

So, the breaking of symmetry implies a sudden change of laws; it is considered as a transition period by physicists, and a revolution for all those who study social change. For science, many other properties of material, all in fact, are created within such transitions period that mark the history of the Universe.

All of this means that scientists are aware of the miraculous phenomena sentences "God said, God made, God created» when they notice some phenomena containing missing links in the Universe. So, they have replaced that by the qualification "symmetry breaking" which means the brutal, even spontaneous transition passage to a state in which a new parameter that did not previously exist come into apparition. This idea is confirmed by the extract below :

"The discoveries in physics over the past decades have led us to give a great importance to the concept of "breaking symmetry". The evolution of the universe from its birth is seen as a succession of breaking symmetries. When they arise from the Big Bang, the universe is symmetrical and without structure. As the time goes by, it cools, it breaks one symmetry after another, and thus allows the appearance of the more and more differentiated structure. The phenomenon of life itself naturally takes place in this way. Life is the breaking of symmetry, too. (Cfr : F. J. Dyson, Infinite in all Directions, Harper and Row, 1988).

As far as creation is concerned, the scientists have noticed the discontinuous phenomena which give birth to other phenomena. But, they have not yet found the causal effect. In the scheme of the biblical creation, the spontaneous breaking of symmetry took place in the areas where Moses uttered "God said, said made or God created".

If science itself cannot find answers to these phenomena, during that time, to whom Moses would have attributed it to being a prophet and scholar?

To God of course. But, the scientists prefer to attribute that to the coincidence of nature. They laugh at the believers while talking about "God said, God made, God created", but they want to convince the believers to accept their thought about the spontaneous breaking of symmetries.

The most specific example about the breaking symmetry in the animal kingdom concerns the case of man and monkey. Indeed, the monkey is often scientifically considered as being the ancestor of man. Now, what we see is that there has been the breaking symmetry that has made us shift from the form of an ape to that of a human being, having all the capacity that allows human to do whatever they think of. Let us keep in mind that, here, the lioness uses her mouth to carry her children, which sometimes causes their deaths, while the man has arms which allow him to adjust well to what he is wearing.

Biblically speaking, after the creation of animals, God took a break. And finally He said : «let us create man like us as our image". We can see that what is going to be created or to form again will be different from the previous one. So, between the first creature and the following one, the scientific analysis have revealed that there is the breaking of symmetry, sometimes called in biology as the accidental mutational of genes.

Section 4 : Parallelisms between standard, astronomical and biblical times

4.1. Standard time

Our time, that is, the standard times use as a reference the revolution of the Earth around the Sun and the rotation of the Earth on its axis. The rotation of the Earth around itself lasts twenty-four hours; which is the equivalent of one day. And the revolution of the Earth around the sun lasts about an average of 365 days that referred to the duration of one year. The latter is known as translation. So when we talk about days and years, everyone takes those standard times as a reference. But concerning the biblical and astronomical time, these values are completely irrelevant.

4.2. **Astronomical time**

The scales of time in the Bible and those of distance in astronomy have almost nothing in common with the daily lives of people on earth. In astronomy, the references are made comparing to the distances. The stars are so far apart from each other so that their meters and kilometers of distancing become unsuitable for expressing them. For this reason, the astronauts have chosen several reference units such as the average distance from the earth to the Sun which they call the astronomical unit (A.U). This distance can count to something like 150,000,000 (one hundred and fifty million) of kilometers.

– A.U. is the reduced distance to one unit between the Earth and the Sun. If 2 stars are separated from one another by 300,000,000 km, we will say that they are separated by a distance of 2 A.U. Thus, Pluto which was the most moving planet in the solar system may sometimes reach an average distance from the sun of 7,395,000,000 km. Dividing this figure by 150,000,000 km, we get 49.3 A.U; which is much easier to count. So, we say the distance between Pluto and the Sun is about 49.3A.U.

But the astronomers did not limit themselves to this. When they began to investigate the space beyond the limits of the solar system, they discovered a so gigantic abyss that even the astronomical units (AUs) became unable far too small as a scale of measurement for that.

For example, among about more than a billion stars already discovered by large telescopes and satellites, the closest called Alpha Centauri is far from the earth at an average distance of 40,398,000,000,000 kilometers (40.398 billion km); a long figure to be written, and is equal to 269,320 A.U.

Thus to evaluate such a distance, astronomers have taken as a reference the speed of light in the vacuum which is about 300,000 km/s or 7 times around the earth in a single second. This is how the appearance of light always seems spontaneous to us. According to special relativity, this speed is a limit that cannot be exceeded by any object with mass. This light consists of waves called photons.

However, this measurement is not suitable for certain distances either, so scientists have chosen as a unit the light-year (al), which is the distance travelled by light in the vacuum during one year. If we do the calculation, one light-year is equal to :

1 al = 300,000 km x 60 (1 minute) x 60 (1 hour) x 24 (1 day) x 365 (1 year) = 9,460,800,000,000 km or approximately 9,461 billion km. So a light year equals to 9.461 billion kilometers.

The Sun is located at about 150,000,000 km from the earth. By dividing this figure by 300,000 km / s with that of the speed of its light, we will have 500 seconds, or something like 8 min 20 s. The sunlight takes 8min20s to reach the earth. To convert this time to km, we multiply it again by 300,000 km / s. This is the way one would proceed to calculate the distance of faraway stars.

Beyond the light year, there is still the parsec. A parsec (symbol "pc") equals 3.08,1013 km, or 3.26 light-years or 206,265 astrono-metric units (AU).

But, just after the Big-bang, during the times that followed, called the inflationary period, events took place at a speed that exceeded the speed of light and of all understanding according to the standard scientific model. The Big Bang Theory seeks to describe the physical state of the Universe during the primordial explosion or expansion. But the models shocked against a time barrier so far, up to now insurmountable set at 10^{-43} (10 exponents -43) seconds after the Big Bang. Below this limit, called Planck's time, the classical theory of gravitation and all physical laws cease to be valid, according to science until then.

For example, during the ultra-short period of time (from 10^{-35} to 10^{-32} seconds), according to science, the volume of the Universe increased in an unimaginable way (by a factor of 10^{27} or 10^{50}, depending on the theories) while in the following 15 billion years its volume increased by only a factor 109.

(Cfr : Big-bang, "Microsoft® Encarta® 2009 [DVD] theory. Micro-soft Corporation, 2008).

As we said, we cited these examples for educational purposes. For us, the events that took place during Planck's time and inflationary time are to be reviewed and we are going to come back to them when we will talk about the theory of inflation versus compared to the theory of the Universal Cosmological Model in Volume II.

4.3. **Biblical times**

As for the Bible, we have said that the notion of time is different from standard, terrestrial and astronomical time. The days correspond to the evolutionary periods and these do not all have the same lengths according to 2 P3 : 8. Let's read :

– Pet*3 : 8 "But there is one thing, beloved, which you must not ignore, and that is before the Lord a day is like a thousand years, and a thousand years are like a day."*

For another down-to-earth example, consider the case of a woman who stayed at the North Pole and testifies that her baby stood up for a day. For us who live between the tropics, this may seem odd to us. However, there are areas in the North Pole, particularly in the Antarctic and Arctic Polar Circles where the alternation of day and night lasts 6 months and not 12 hours. So, looking back in time and space, this woman told the truth because where she gave birth a day is 6 months long compared to us who are between the tropics. And today, each planet has its day length relative to its rotation.

The story of creation was rejected by science. No one had the idea of putting it back in time and space to understand this language that comes to us from ancient times. Fortunately, the Principle of Creation, chapter of the Divine Principle of the Moon Couple, has cleared up the ambiguity

Moses, through the biblical writings, also used a measure to indicate time in creation. The Rev. Dr. SUN MYUNG MOON was the first in the Creation Principle, branch of the Divine Principle, to disclose that the days in the biblical creation were periods, while until then everyone believed in a simultaneous occurrence of events dictated by the Word of God. So, a biblical day is the time it takes for an event to evolve from its conception (by God), until its realization - what we call the period of maturity or morning according to the Bible.

Indeed, before the Big Bang and in the first moments after the Big Bang or the separation of the waters from the biblical waters, there was no star, nor any star. That is, there was neither day nor night. This makes us unable to determine the durations of the events that have taken place.

For example, the 380,000 years which represent the time between the Big bang until the recombination and, which manifest itself by the reappearance of the light as determined by the scientists, were obtained by simulation, i.e. that is, creation of an apparent act which does not correspond to a real transaction. It is the same for all the events included the inflationary time. During this time, there was no alternation of day and night. And when there is no alternation of day and night, the end of one evolution to pass into the other marks a period. And that's what Moses did in the Bible, calling it a day.

While astronomers used certain material references to determine time, Moses was talking about creation in the universe instead. In other words at the time when there was no reference to indicate the time. Thus, we understand why he used it as a reference "day" ; the time when an ideal was conceived until its materialization. The appropriate scientific term is the cycle of evolution. In this way, we immediately understand that these days did not have the same duration.

Hence this scientific conclusion :

"*... time varies according to quantum processes and therefore there is no such thing as an even abstract universal time (like that of our current life) which can define variations between two states of a quantum process*". (Cfr : Claude Sarrailh, the secret of life, ed. Librinova, 2018, P.12).

For example, the first period called the first day spoke of the time when the universe was empty and dark until the appearance of light (Ge 1 : 1-3). Through this book, we will discover that it was a much longer time than the second day when it is written that "God divided the water from the waters" or Big-bang and which only lasted for the space of 10^{-n} second.

Here is a scientific illustration of what has happened immediately just after the "separation of the water from the waters" or Big-bang

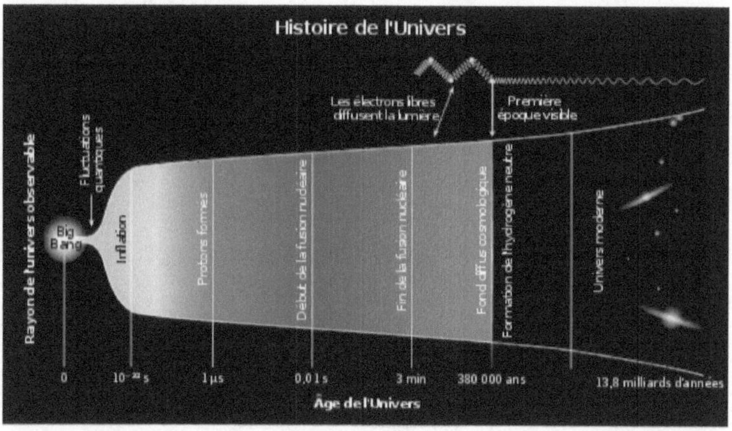

"The epoch of hydrogen recombination corresponds to that of the emission of the cosmic diffuse background, at the border between the purple and black areas of the diagram." Source : (https://upload.wikimedia.org/wikipedia/8/84/History_of_the_Universe_crop_fr.svg.png).

And finally, there are still two speeds that are above all others. It is that of the Divine Himself which allows Him to be permanent on all points of the cosmos and that of the mind which acts simultaneously with thought, to which physics has no access. Nevertheless, since this book is the first to talk about it scientifically, I was able to calculate the minimum variables within the framework of the Universal Cosmological Model in Volume II.

Chapter III : Cosmology according to the biblical model

The Biblical Model is the story of creation as narrated by the Bible in the first two chapters of Genesis. This book is intended to be a scientific restitution of this model, using the current development of the biblical diagram of creation, as restored and analyzed in this book, entitled the "field of Universal Cosmological Model". This scheme has never captured the attention of scientists, and it is rarely chosen as the topic subject of sermons. In bookstores until the publication of this book, there was none which has tackled the same theme.

The reason is very simple : this story is part of the scientific-religious field. And in order to fathom it, in addition to biblical knowledge, one needs knowledge of the preliminary notions of cosmology and the Principle of Creation contained in the book of the Divine Principle by Rev. Dr. Sun Myung Moon. Being among the first person in the world to have focused to learn all the three knowledge at once until the apparition of this book, it is understandable the reason why I am the first to deal with this theme. This proves that the comprehension of the Genesis story about the creation of the universe in a scientific way is not as easy as people used to think, except after reading this book.

Despite my research and knowledge, I have not reached a full understanding of this scientific restitution, and I am not yet able to express it perfectly. This consideration concerns especially the Before-Big-bang, because the After-Big-bang is widely analyzed by science. Unfortunately, it took a bad start by making the Big-bang the point of origin of the universe and by excluding God.

What I present in this book, namely cosmology according to the Universal Cosmological Model is only an introduction to open the door to the masters of this discipline who are astrophysicists, cosmologists, astronomers, planetologists, physicists, astrochemists and all those who devote themselves to the study of the origin of creation. This is an introduction to allow them to enter and develop it in a scientific way. However, no one will be able to truly fathom the keys of the Universe in a perfect way. Because even if we have within us the spirit of him who witnessed these things, namely Christ (Jn 1 : 2), but we ourselves are only the result.

In addition, in science we say that an answer gives rise to other questions. As long as we have not yet arrived at the origin, which is God, we will always have questions.

There are scholars of astrophysics and exegetes of the word of God, but, before the appearance of this book, I was the only one in mankind to be interested in these two disciplines at the same time. So I discovered that the history of creation according to the Bible is scientifically logical, richer and more coherent than any model that science has offered us. It is this discovery that I pass on to humanity that makes me a self-taught cosmologist.

Thus, branches such as the preliminary notion of cosmology, the Divine Principle of the Moon couple, and the biblical schema on creation scientifically restored in the Universal Cosmological Model must be included in all theological, philosophical and those of certain scientific field. Otherwise, the philosophical reasons given by the latter will always remain partial.

Just as everything astrophysicists see as part of effects, which are the subject of their science, they must include cause as complementary knowledge in their discipline. And the cause is indeed the teaching of the Divine Principle of the Moon couple, as the philosophical basis of their teaching or knowledge. It is only through this teaching that they will be able to solve many puzzles. This concern responds to the word of Jesus who said to his disciples : "I am speaking to you of earthly things and you do not believe, how will you believe when I tell you of heavenly things ?" (Jn 3:12).

This promise of Jesus is fulfilled to this day by the MOON couple. All that remains are for the intellectual leaders to capture it and involve it in their discipline. Therefore, no one can better explain the Universal Cosmological Model without the elementary knowledge of the Word of God, of cosmology and of the Principle of creation contained in the book of the Divine Principle of the Moon couple.

The knowledge of the teachings of the Moon couple and the preliminary notions in cosmology allowed me not only to clarify the history of biblical creation but also the ancient models drawn and exploited by the various sects, including esoteric sects. It has also allowed me to offer a clear explanation of the formation of man according to the Quran and to make certain hypotheses in cosmology on which science will decide later. Everything is recorded in this first part.

Section 1 : The Universal Cosmological Model

1. Presentation

The Universal Cosmological Model is the analytical restitution of the biblical scheme of creation. The word "analytical" is defined in philosophy as an inference from principles based on experience and observation (Dicos encarta). It is thus understood that this restitution has nothing to do with a sermon based on faith. All our analyses are supported by the experiences, observations and conclusions that were made by scholars in different fields. Others will be in the centuries to come.

We have said that the biblical model is the creation story as written in the Bible. Our restitution can be criticized by anyone, as long as he does not touch the word "Bible or biblical model" for respecting the latter name and not to tarnish its sanctity. This is how we have entitled our restitution, which can be criticised, "the Universal Cosmological Model". Thus, in making our scientific commentary on the biblical creation story, we enter directly into the Universal Cosmological Model.

My only addition compared to the analyzes which will come to support the biblical diagram of the creation later, is that of having opened the door, and to have it scientifically approved, even if the arguments are recorded in this simple reading book. This is my only way to get in everywhere and teach humanity.

My great contribution is also to have clearly explained the words, names and phrases that made the diagram of the biblical creation seem like a magician's game with phrases like "God said, God made, God created".

2. The basic statement of the Universal Cosmological Model

Thus, the Universal Cosmological Model, on which all analyses will be based, reads as follows :

The Universe was formless and empty. Given, on the one hand, an energy defined as darkness by the Bible in the abyss, and on the other hand, the spirit of God or a wind from God on the original material called "waters", there was the production of a radiation defined as light. From the beginning of the Universe in nothingness until the first light or radiation was produced, this is the period that the Bible calls 'day' (first day), a period that lasted for a time that could not be quantified until then.

The Bible further specified that this new energy or radiation had an effect on darkness. It came to counteract it, as well as the invisible material that was already at work in the early universe. This is expressed in the sentence "God separated the light from the darkness". And to show that these two energies would henceforth accompany the physical destiny of the Universe, God called darkness "night" and light or new radiation "day", for lack of the appropriate scientific words at that time.

The original material or "waters" according to the biblical term, continuing its warming cycle after the production of light (light be and light was), had a burst, an expansion or an expulsion into the abyss and that is what science calls "Big-bang" and the Bible calls "separation of the waters from the waters. This evolutionary period of material warming (waters) from the production of light or first radiation, to the Big-bang, was described by the Bible as the second day. And then came the third day and the fourth day which marked the beginning of the post-Big Bang period which were periods of the creation of the stars from the same material or waters according to the Bible.

Then came the fifth biblical day, the period that witnessed the appearance of living beings from microscopic underwater beings to birds.

And at the end or on the sixth biblical day, there was the appearance of mammals and of humans. (Gen 1:1-30; Gen 2:1-7).

This constitutes the scientific restitution of the writings of Moses dated since 1500 B.C. This is the basic statement of the scheme *of creation according to the Universal Cosmological Model.*

These are exposed in a new way for astrophysicists, cosmologists, astronomers, theologians, philosophers and my other readers, the Biblical Model on which all the other Models that will support it should be built.

2.1. **The basis of the statement**

Indeed, scientists have understood that it is through the changes in temperature that they can understand how the observable universe appeared from cooling to high heat or vice versa, on original particles. This rise in temperature was also observed in the Bible through the sentence "God said, let there be light! And there was light "under the power of the spirit of God or a wind of God on the waters which represent the original elements.

As a laboratory for simulations to understand what happened at the beginning of the Universe, scientists built the Large Hadron Collider (LHC) straddling Switzerland and France, managed by CERN (European Organization for Nuclear Research).

In this gas pedal, two beams of high-energy particles travel at a speed close to the speed of light before colliding and creating elements by emitting strong light and energy. Through these processes, researchers are trying to demystify the Universe and understand how elements were formed. The verses of Ge1 : 2-3 "the spirit of God (energy) moving over the waters (original matter), light was (sprang)", became scientifically indisputable and brought this biblical scheme from cosmogony to cosmology.

3. **original elements, original particles, original materials**

Why do we use the expressions : "original element", "original particle", or "original material" instead of elementary particle, or fundamental particle? Indeed, in science we mean by elementary particle, or fundamental particle, a particle whose composition we do not know : we do not know if it is made up of other smaller particles.

But, since elementary particles include the "fundamental fermions" (quarks, leptons, and their antiparticles, antiquarks and antileptons) that compose the material and antimaterial, as well as bosons (gauge and Higgs bosons) which are vectors of forces ,and play a mediating role in the elementary interactions among fermions (Cfr : Wikipedia; elementary particles), and not knowing with precision what was the matter before the Big-bang, from which these elements come from, we call them either the original elements, or the original particles, or still the original matters, in reference to God who is the origin of all things or the origin of origins according to the oriental philosophy. This reference to God has nothing to do with a belief just as the names of the planets of the solar system have their origins in Greek mythology.

In addition, elementary particles are the smallest physical objects of which matter and the forces of the universe are made up. They are packets of energy characterized mainly by mass, and other numbers such as electric charge. They acquire these properties only from the original particles, emanating from the dual characteristics of God. (Cfr : Presentation of the Divine Principle : ed. Culture et Paix, Paris, 2013, Pp 26-43).

Section 2 : The biblical creation scheme

2.1. The stages of the universe's evolution according to the Bible

Indeed, no theologian till today on has had the idea of reasoning beyond the biblical writings. Thus, the verses below make us think of the different sentences about the creation history of the universe according to the Bible, also certified by science.

– Ge *1 : 1-2 "In the beginning, when God created the cosmos, (the heavens and the earth). The earth* (the universe) *was formless and empty ".*

– Ge 1 : 2 *"There was darkness in the surface of the abyss* (the first deformation of the fabric of space-time) *in the axis of the universe".*

– Ge 1 : 2 *"The power of God was moving over the water* (original materials before the Big-bang in the deep abyss)"

– Ge 1 : 3 *"God commanded, 'Let there be light, and light appeared';* (consequence of the Spirit of God moving over the water or over the original material) (Ge 1 : 2)

– Ge 1 : 4 *"God separated the light from the darkness"*

– Ge 1 : 7 *"Separating the waters from the water"* or Big bang according to science (1) and expanse creation (2).

– Ge 1 : 9 *"From the waters below* (1) *the dry* (2) (or the earth) *appeared"*

– Ge 1 : 14 *"From the waters above* (1) *there was creation of lights* (2)"

All these phenomena or developments take place with the same original materials, so-called as "water" according to the Bible. They are activated by the same energies, also so-called "darkness" and "the spirit of God». These are phenomena that make the Universe isotropic (isotropy : The Universe has the same properties in all directions) and homogeneous (Homogeneity : the physical laws are the same in each place).

This tells us that the Universe does indeed have an origin and a starting point as it is described by the Bible. This is what is reproduced scientifically in this book under the name of "The Universal Cosmological Model" and which gobbles the Standard Model, while correcting the other models and scenarios imagined by scientists.

Indeed, the development of human models and their conclusion can only find their reason in the "Universal Cosmological Model".

In addition to its steps, I recommend my readers to follow the first five minutes of the scientific documentary entitled : "Hubble from Space : The Work of Light" on You tube.

In fact, without the director knowing it, this documentary retraces the scenario of the reappearance of light 380,000 years after the Big Bang. But, it said nothing about the appearance of this light since its beginning. This light, according to the Bible, did well and truly appear before the Big Bang or "the separation of the water from the waters." By the way, this scenario is similar to what is described in Ge 1 : 1-3.

Let's read :

Gen 1:1 In the beginning God created the heavens and the earth. 2 The earth was formless and void, and darkness was over the surface of the deep, and the Spirit of God moved over the waters. 3 God said, "Let there be light. And there was light.

In the above-mentioned documentary, the gravity plays the role of the spirit of God whereas the elementary particles play that of the "biblical waters", then, the concentration of particles creating space-time and making the universe an abyss. That is how the appearance of light came. Here is a proof to support in force the Universal Cosmological Model.

Why "in force»? Because it is a scientific proof which has nothing to do with this book and which was conceived out of any idea written in the Bible. In addition, the director talks about a scenario after Big-bang, whereas for us, it is a scenario similar to what took place before the Big-bang until the emergence of the light which is written as follows : "God recommended : let there be light!" And there was light" (Ge 1 : 3).

We have already explained that the phrase "God says" is to demonstrate that this was part of His will. But scientifically, it was the gravitational result on the original matter known biblically by the phrase : "the Spirit of God moved over the waters" (Gen 1:2) as we can see in the above mentioned.

2.2 Illustrative tableau on biblical and scientific cosmology

BIBLICAL PERIODS	CREATION	KEY WORDS	SCIENCE
SPACE -TIME	Ge 1 :1Inthebeginning,God created the heavens and the earth.	-Heavens (Spiritual World), - Earth (Physical world)	Before the Big-bang
DAY 1	Ge 1 :2-3 « 2 The earth was formless and empty; there was darkness on the face of the deep, and the Spirit of God was moving over the water. 3 God said: "Let there be light ! And the light appeared"	— - Empty –Darkness –Abyss -Spirit of God – Waters Lights	
DAY 2	Ge 1 :6 God said, "Let there be an expanse between the waters, and let it divide the waters from the waters."	-Extent -Water separation	The BIG-BANG
DAY 3	Ge 1 :9-11 God commanded, "Let there be a dome to divide the water which are below the heavens be gathered together in one place, and dryness appear. And so it was.10 God called the dry land earth, and he called the cluster of waters seas. "11 Then God said," Let the earth bring forth greenery "	— — Seas – Earth -Herbs (Molecules of life)	-Formation of elements up to galaxies (stars,lights and planets)

DAY 4	Ge 1 :14 God said: "Let there be lights in the expanse of the sky to separate the day from the night; let them be signs to mark the epochs, the days and the years"	-luminaries (Stars)	
DAY 5	Ge 1 :20 God said, "Let the water bring forth living animals in abundance, and let birds fly over the earth to the expanse of heaven."	-marine beings (Plants) - seabirds (plants, bacteria, Pisces) - flying beings	-marine beings (Plants) -seabirds (plants, bacteria, Pisces) -flying beings

Chapter IV :
The physics of the creation

In the Bible, it is written that originally "the earth was formless and empty" which is the shape of the Universe at its beginning. How come from a vacuum did the first elementary particles appear ? Such question is the paradox of Phanzu, to which the physicists about the creation of the universe try to answer. Thus, the physics of the creation encompasses all the sub branches of physics or all the thoughts of the physicists who reasoned on the transformation of the world which goes from the emptiness to the appearance of the first particles.

Many astrophysicists have worked on this subject in disparate ways. But by bringing their thoughts together, we do end up with argument that the universe had a beginning as the Bible has stated.

As Volume I of this book is a synonym for the introduction of the Universal Cosmological Model, talking about it here would affect readers. Nevertheless, we want to talk about what we have retained from the two main branches namely the Principle of creation and quantum mechanics. However, quantum mechanics being a purely scientific branch, we will only talk about it in Volume II.

Section1 : The Principle of the creation of Rev. Dr. Sun Myung Moon

1. Presentation

The Principle of Creation of Rev. Dr. Sun Myung Moon, recounted in The Divine Principle book, is "the universe's only manual" for men. Until the advent of Rev. Dr. Sun Myung Moon, man found himself without a true parent who could teach him why the universe was created, what God's ideal was, and why that ideal was not achieved. Apart from the Principle of creation which is only exploited in this Volume, we find in the same book about the Divine Principle, the origin of the falling down into sin of the mankind, and the providence of God for his restoration.

Many religious will not appreciate talking about the Principle of Creation of Rev. Dr. Sun Myung Moon as the only way to use the universe. Indeed, all the main figures of the world great religious confessions have focused on the relationship between man and God, man and his neighbor, and how to live according to the will of God. However, their messages were only partial truths. This is why they always prophesied concerning the coming of a great truth at the end of time. Even Christianity is included among them. Jesus even reminded people about it.

Let us read : Jn*14:26 "But the Comforter, the Holy Spirit, whom the Father will send in my name, will teach you all things, and will remind you of all that I have said to you".*

Here we are dealing with the Holy Spirit who can only teach through a person and not with the Holy Spirit whose Christians obtain in Baptism. Moreover, a revelation is not a teaching in itself solely. He only becomes so through man.

Saint Paul also returned to these words of Jesus through these verses :

I Cor*13 : 9-10 "For we know in part, and we prophesy in part, but when that which is perfect shall come, that which is partial shall be done away with."*

He confirms that our current knowledge about religions and prophecies are only partial. Hence the search for more truth is given to everyone. However, the book of Revelation or Revelations which speak of what should happen in the future, confirms everything that is said above through the following verse :

Rev *14 : 6 "Then I saw another angel flying in midair, and he had the eternal gospel to proclaim to those who live on the earth—to every nation, tribe, every language, and people.*

The book of Revelation announces only some future prophecies. Saint Paul has spoken about the partial knowledge that is taught to us now through all current sacred books, but the apostle John in his book of Revelation announces the coming of an everlasting Gospel which will be taught to everyone without exception. This teaching is going to create the unification of the Word of God and the universal knowledge for everyone being Muslims, Christians, Buddhists, etc.

The Principle of Creation, branch of the book "Divine Principle" written by Rev. Dr. SM MOON, and the book "THE VERSETS OF THE SCHOLARS" Volume I, now for the first time constitutes the fundamental source of inspiration for researchers to integrate from now on into the scientific understanding of God, as well as on the origins of the universe up to the appearance of the mankind in the second.

Indeed, as in the time of Jesus when he regretted to find himself inside a humanity which was not prepared to receive great things as confirmed by John 16:12 "I still have many things for you. say, but you can't wear it now," the Rev. Dr. Sun Myung Moon, who despite his efforts to bring scientists together on different occasions, ignored these works, bewildered by the biased judgments of the media. The latter treated him with all evils as they did with Jesus : brainwashers, tradesman, etc. Is it not written that Christ will return with power and glory, in opposition to his first coming where even the fox was in a more envious situation than him (Mt 8:20) ? We will come back to this in other volumes which are going to reveal all the symbolisms about the second coming of Christ.

Knowing better the universe and its principles, and because he was able to explain it better to scholars without using parables and symbolisms, but with an appropriate scientific language, the Rev. Dr. Sun Myung Moon was obliged to study engineering in Japan. These conclusions which better explain God, the origin and eternities of the universe, are found in the Principle of creation, chapter of the Divine Principle, and at his many works. And as I said, it is on the basis of these works that I came to design the Universal Cosmological Model. However, this model has nothing to do with the doctrine of the Church of Rev. Dr. Sun Myung Moon. Rather, his teaching served me as a knowledge base.

Thus, all theories about the origins of the universe, beaten without knowledge of the principle of creation, branch of the divine principle of Rev. Dr. S.M. Moon, will generate a lot of puzzles and are doomed to destruction one day. Moreover, it is only by studying the Principle of Creation that the phrase "God says", which is the foundation of all things, changes from the level of belief to the level of reality. At least, for those scholars who agree themselves when someone talks to them about energy, particle, mass or material, compared to theologians and philosophers

The Principle of Creation, the branch of the Divine Principle written by the Rev. Dr. Sun Myung Moon, although it is true, but it has been unnoticed within the scientific world, because it is taught as a doctrinal teaching of a church, namely the Unification Church. In addition, it is recorded in a religious book. I have already said before that all scientific truth, recorded in a religious or sacred book, is like a hidden treasure out for the public. This is why some of the verses quoted in this book will amaze scientists. They are going to discover that these phenomena were already unveiled before their scientific predictions and observations being done.

Today science is troubled with many questions about cosmology because it only analyzes the physical or visible aspect of the universe and draws inferences from it, while ignoring the spiritual or causal aspect in the one hand. The Principle of Creation, on the other hand, integrates the causal aspect and seeks to explain the functioning of the universe from the characteristics of God himself.

Stepping into the infinitely small, Niels Borg, one of the fathers of quantum physics said; "Anyone who is not shocked by quantum theory does not understand it." It was this shock that I also felt as I studied Rev.'s Principle of Creation. Dr. Sun Myung Moon who endeavored to convey how the macrocosm is a manifestation of the microcosm and vice versa. So, I will also say that anyone who has not felt this shock while studying the Principle of creation will never understand the divine Principle.

Like the Bible and the Koran, the Principle of Creation, a branch of the divine Principle, is considered by the followers of Rev. Dr. Sun Myung Moon as God's revelation revealed to their masters. We should respect their points of views, and also offer the same respect to their messages as we reserve to the Bible and the Koran.

Since the creation of the universe, the content of these Principles is the only message that unveils the ideal of God in creating the mankind. Thus, in order not to make this content as an object of discussion in the scientific world, we have selected only 4 points of the Principle of creation which are going to interest the scientists, and that will be a part of the "Physics of Creation", just like quantum physics. Thanks to these four features of the Principle of Creation, several riddles of the universe will be solved by scientists, as Ms. Moon said in one of these recent speeches :

"When you take into consideration what is taught in the Divine Principle and start to apply it in the different areas, you are going to discover more, and you will be able to explain a lot of things (enigma)."

2. The four features of the Principle of creation

The four features of the Principle of creation; branch of the Divine Principle book of Rev. Sun Myung Moon who will be part of the physics of creation and will now be the subjects of study are as follows :

- The dual characteristics of God, which explains how the universe functions;
- The Universal First Energy : which describes the source of energy and its extension in the universe;
- The period of growth of the created beings, which reveals that the biblical days are the periods;
- The spiritual and physical world who's the Rev. Dr. Sun Myung was the only one to describe the relations that unit both and form the cosmos.

All these four points of the Principle of creation are now shifted towards science for their deepening into the "Physics of creation".

It is the same thing that I did concerning the Bible and

other sacred books related to science, which I restore under the name of the "Universal Cosmological Model" in order to respect the names of these sacred books. So, when we are speaking about the physics of creation and the Universal Cosmological Model, we avoid any bad judgment that can make these sacred writings like the Divine Principle, the Bible or other sacred books doubtful.

I myself bear the responsibility for the judgments of the selected writings or verses. So, I offer you these two definitions :

- The Universal Cosmological Model is the scientific restitution of the story of the biblical creation. Its scope begins when the biblical creation narration speaks about " the shapeless and empty earth" which shows the beginning of the universe originally till the last stage when God created one man and one woman.

- The Physics of Creation encompasses all the branches of physics or all the thoughts of physicists, religious or philosophers who have developed their thoughts about the world creation from emptiness to the appearance of the first particles.

- The book "THE VERSETS OF THE Scholars" as for itself, extends its field of action until the second advent of Christ which will be developed in the other Tomes.

We understand that to be able to write the full book on the entire cycle of the universe, you have to be a multidisciplinary scholar like Moses the scholar of scholars, a Hero like Rev. Dr. Sun Myung Moon who was not afraid to enter into to the hell to take out from the ex-angel of light all the truth about his downfall, and an exegete like Flavien PHANZU MWAKA. If all three of these characters managed and succeeded with divine assistance, then why not you?

So, let's see an overview of the points of the Principle

of creation, taken within the framework of the physics of creation which has just enriched science.

1.2.1. Dual characteristics of God

The universe is being the visible extension of the invisible God, we can better understand its functioning only through the common and visible characteristics of creation as explained in the Principle of creation. These characteristics find their source in the invisible God manifested through creation as confirmed by the following verse :

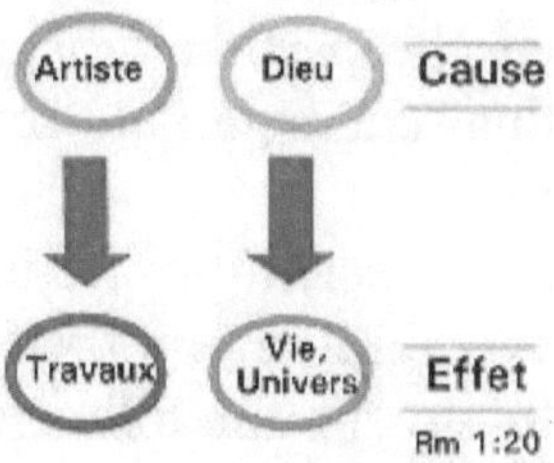

Rm1 : 20 *"Indeed, the invisible perfections of God, his eternal power and his divinity, are seen as with the naked eye, since the creation of the world, when one considers them in his works. They are therefore inexcusable"*

Source : The Divine Principles, level 4, p.13

Saint Paul's great wisdom in this verse (Rom 1:20) reminds us that anyone who says he does not know God is not analyzing the creation and himself well. The Rev. Dr. Sun Myung Moon was the first to point it out in the Principle of Creation as the "Dual Characteristics of God". During his researches, he came to the conclusion that all beings are made up of the following dual characteristics.

– All materials, from particles to humans are made up of an invisible entity and an external form called mass by the scientists according to the following table :

CREATION	Invisibleentity (internal)	Mass [external form]
Male	Heart [soul] and Spirit	Body
Animals	Animal spirit	Body
Plants	Plant spirit	Body
Molecules	Inherent nature	Energy/material
Atoms	Inherent nature	Energy/material
Particules	Inherent nature	Energy/material

As Saint Paul has pointed it out, by examining all these characteristics that are universally present in the various elements of creation, we come to understand now the nature of God : He is the First Cause of the whole universe as well as the partner-subject himself who has the harmonious dual characteristics of inner nature and original outer form.

Everything that is scientifically unexplained in his creatures finds its causation in their invisible entities. The Bible describes this by the following verse "Then the Lord God formed a man[c] from the dust of the ground and breathed into his nostrils the breath of life, and the man became a living being" [Ge 2 : 7]. The dust of the earth representing the mass, and, the invisible entity is represented by the breath of life. A detailed explanation is contained in the book "Presentation of the divine principle : ed. Culture and Peace, Paris, 2013, Pp 21-25".

– All materials, from particles to humans, are made up of a positive entity and a negative entity according to the following tableau :

CREATION	Positive entity	Negative entity
Man	Male	Female
Animals	Male	Female
Plants	Etamine	Pistil
Molecules	Cation	Anion
Atoms	Positive	negative
Particles	Positive	negative

According to the principle of creation, the branch of the divine principle of Rev. Dr. Sun Myung Moon, we read this :

"In the nature, we find that within all creation there is a reciprocal relationship between positivity and negativity. For example, atoms are formed from the reciprocal relationship between positive elements and negative elements. The atoms themselves, composed of the particles, have positive or negative characteristics. Based on such characteristics, two or more atoms maintain a reciprocal relationship, forming molecules. All plants and animals exist and are renewed by virtue of myriads of interrelated inner and outer relationships of positive and negative elements. ". Humanity is made up of men and women and human society exists, multiplies and develops through reciprocal relationships between men and women "-(Cfr : Divine principle, level 4, Pp.15-17)

While mixing these tableaus, the Rev. Dr. Sun Myung Moon came to the conclusion that the relationship between positivity and negativity of beings is similar to that which exists between the invisible entity of each body and its mass. Thus, the creation is made in such a way that everything exists by a reciprocal relation of positivity and negativity. From that several laws flow that science has formulated as Coulomb's law which says that "two charges of opposing signs attract" and tend to return to unity, I added. By this, I also conclude that gravity is one of the consequences of these characteristics.

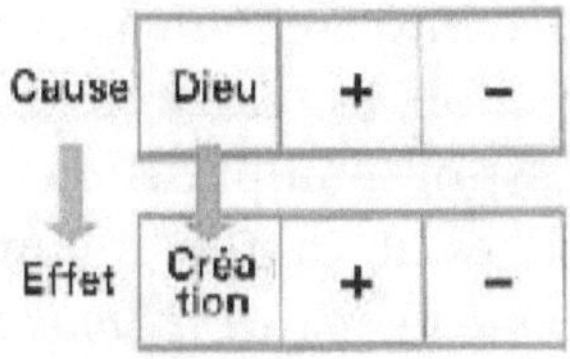

Source : The Divine Principles, level 4, p.17

Thus, having discovered these unseen dual characteristics in creation which in fact is only the effect, the Rev. Dr. Sun Myung Moon concludes that the origin of all these should come from the cause, that is to say, the Creator who is God. Therefore, the universe is controlled under the laws and principles of God himself who is the embodiment of these same dual characteristics : invisible-mass, positivity-negativity, masculinity-femininity. It should be emphasized that the invisibility of God and his outward form do not exist as independent entities, but in harmonious reciprocal relationship.

From the light of this understanding about dual characteristics, the relationship between God and the universe can be summed up as follows : God is the invisible subject partner and the universe as a whole is in the position of a substantial object partner with God. The universe is made up of individual incarnations of truthiness; each of them is a unique manifestation of the dual characteristics of God either in image or symbol, according to the Principle of creation. (Cfr : Presentation of the Divine Principle : ed. Culture et Paix, Paris, 2013, Pp 27).

All of these characteristics are being united and existing in harmony in God are called original dual characteristics.

As they are projected into creation, they tend, according to the Principle of creation, to return to unity by creating on their way the process of multiplication and development.

The Bible describes it in the following verses : "*So God created mankind in his own image, in the image of God he created them; male and female he created them.*" (Ge1 : 27). Man and woman have inside them an invisible spirit and an outward form (mass). In addition, the man represents positivity and the woman negativity. The source of all this is God according to the above verse. All these characteristics will also result in the union of man and woman, and Moses brought it out as a law in a figurative sense : "*That is why a man leaves his father and mother, and is united with his wife, and they become one flesh.*" (Gen. 2:24).

One flesh means the recreation process of the original image, that is, the image of God. It is this law, we have said so far, which governs the whole universe, making it dynamic by rotary and gravitational movements. Unity and unification encompass the physical term known as "fusion". This can be illustrated by the diagram below.

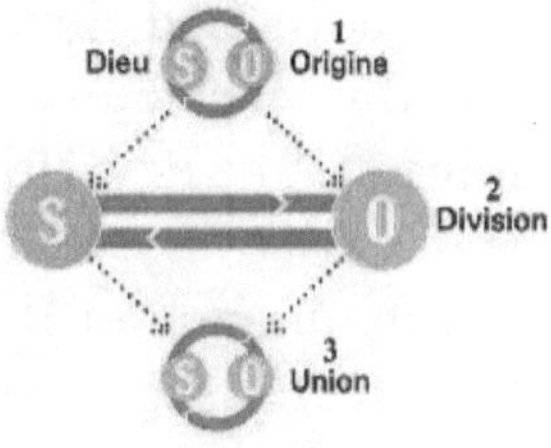

Source : The Divine Principles, level 4, p.24

In this diagram, God, as being the origin of everything, has possesses himself in his invisibility all the dual characteristics represented by S, that is to say, subject as active character ,and O that is to say, Object as passive character, but both work in harmony. During the creation process, the projection of the Subject and the Object is divided, it starts with the particles-antiparticles, virtual as they are. According to Coulomb's law in which two charges of opposite signs attract each other, the subject (particle) and the object (antiparticle) will interact by annihilating each other. When we talk about the Universal Cosmological Model, a new particle that comes from the mixing of the two properties will be created.

Taking into account the collision, the result is going to be as either the new particle has a dominant interaction comparing to the other that has a minus one or vice versa. This will lead to more new combinations or interactions until the creation of the elements.

The study of the dual characteristics of God in creation brings to light what we call "the uniformity of creation." It is on the basis of how this uniformity works that all the world puzzles can be solved. Thus, thanks to the study of the dual characteristics of God that the "theory of everything" can spring.

As we have just seen above, God himself cannot be active without energy. He is thus the source of all the energies which govern the universe. The Rev. Dr. Sun Myung Moon was the first to make this discovery. He transcribed it in the Principle of Creation, the branch of the divine Principle, as we are going to discover it succinctly in the following point.

1.2.2. Universal Prime Energy

To exist as an eternal and active entity, God needs to have the energy. It is this energy contained in God that is the source of all energies. It is called the Original Energy, labelled the "Universal First Energy" by the one who has discovered this notion, namely, the Rev. Dr. Sun Myung Moon. (Cfr : Presentation of the Divine Principle : ed. Culture et Paix, Paris, 2013, Pp 29- 32).

It is the notion that has nothing to do with the church doctrine, it is purely a scientific one. It is this notion that makes the Principle of Creation of Rev. Dr. Sun Myung Moon greater in comparison to all current scientific theories ; quantum physics and quantum mechanics included.

Even though this notion is a part of the general teaching within the church of Rev. Dr. Sun Myung Moon which explains the origin of all things, as an economist, I never quite understood it, and that's why I don't want to deal too long with this point.

Considering the intellectual level of that time, Moses used man as a scenario to state the great laws of the universe that we are going to restore scientifically throughout this book.

As we have said above, according to the Principle of Creation, the UPE (Universal Prime Energy) is the origin of all the energies and all the forces which allow to all created beings to exist. Thanks to the action of the universal primary energy, the subject and object elements of each entity form a common basis and interact according to the diagram below :

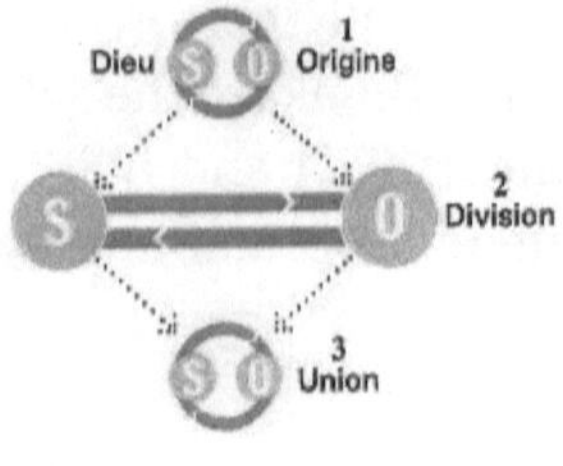

Source : The Divine Principle, level 4, p.24

In the created universe, the dual characteristics that compose each being receive the power of the universal primary energy to establish a common basis. Then they begin a reciprocal action of giving and receiving.

This relationship in turn generates all the forces that the entity needs its existence, multiplication and action. The interaction which thus is generated by these forces is called in the Principle of Creation "reciprocal action of giving and receiving». *(Cfr : Presentation of the Divine Principle : ed. Culture et Paix, Paris, 2013, Pp 29-32).*

By relying on the forces generated by this reciprocal action of giving and receiving according to the Principle of Creation, the dual characteristics establish the basis for their eternal exchange. This is the foundation by which God can exist eternally and generate all the forces necessary to create and master the universe. (*Cfr : Presentation of the Divine Principle : ed. Culture et Paix, Paris, 2013, Pp 30-32).*

As we said, the literal understanding of all this is very complicated and it requires a lot of diagrams. Let us not forget that it is in this point of the dual characteristics of God that all the mysteries are hidden; not only of the Universe, but of beings as well. You just need to understand these concepts and apply them in the entity you are studying. What interested me the most was understanding the origin of rotation, starting with particle spins and what science calls gravity.

Indeed, the spin is a fundamental intrinsic characteristic of elementary particles. The effects of the dual characteristics of God already start from some elementary particles. For example, we notice that an electron only returns to its starting state by a double rotation : this is called a particle of spin ½.

1.2.3 The period of growth for created beings

I have already dealt with this point in the discrepancy between religion and science regarding the notion of time.

Indeed, in the Principle of Creation, the Rev. Dr. SM Moon was the first religious to draw a parallelism between the day of creation in Genesis and verse 2 P 3 : 8 which states : ”*...there is no difference in the Lord's sight between one day and a thousand years, to him the two are the same : a day is like a thousand years and a thousand years are like a day* ”. By this we understand that the days of creation in Genesis were not actually 24 hour days. They were more like periods instead.

The Rev. Dr. S.M. MOON was also the first to be able to explain the biblical language of “... *one evening and one morning,*” which are periods of growth and evolution. The first chapter of Genesis tells us that the universe did not exist instantly, but it was created during six gradual periods. To this, let us read this in the divine Principle :

”Morning” means the period during which the ideal of creation was first realized. Whereas the ”Evening” means the period when God ceased to create. By understanding that the morning came after the evening and night, it means that God considered both of these periods forming only one day. God

indicated what time was required for his ideal of creation to be realized. So, the period from which God started the process of creation up to the realization of his ideal is considered as a period of growth for all created beings, and it is symbolized by night, the period located between evening and morning". (Cfr : Divine principle, level 4, p.32).

1.2.4. The invisible substantial world and the visible substantial world

The invisible world or spiritual world is a theoretical reality according to believers, but a non-existent world for the scientists. The Rev. Dr. Sun Myung MOON is the first to have actually described it during his various lectures, and he defined it in the Divine Principle. The ignorance which still characterizes all mankind concerning this world is due to the lack of knowledge about the teachings of the Divine Principle of Rev. Dr. Sun Myung Moon. That is why this teaching must be one of the branches of the courses in all the theological faculties curriculum without any exception.

For science, as we have said, the cosmos represents the universe taken as a whole, that is to say, from the interstellar space to the organized system. In other words, it is the observable physical space.

According to the principle of creation, the branch of the divine principle of Rev. Dr. Sun Myung Moon, the cosmos represents the whole spiritual world and the universe or physical world together. Man as being composed of two united entities, namely the mind and the body, constitutes the center of the cosmos. The body is composed of the physical elements of the cosmos and the spirit of the spiritual elements of the cosmos, thus man is the microcosm or a cosmos in miniature.

The difference between spiritual entity and physical

entity is that our physical body is limited by time, it cannot transcend the present moment. Our physical body is an ephemeral being, it ages and returns to the earth. Our mind is not limited by time; no time barrier can stop it. He can freely reflect on the past and contemplate the future if he wishes; it is eternal.

According to the Universal Cosmological Model, which makes the teaching of the Principle of Creation its own, the cosmos can only be better understood by studying man. So, why not search in man what is still invisible or mysterious in the knowledge of the universe?

Section 2 : Confirmation of the Principle of creation

2.1. Asteroid RYUGU

A theory can only be accepted in science if it is mathematically proven valid after experimentation or if it is confirmed by observations. What I discovered in 2018 on the asteroid Ryugu almost confirmed what is taught in The Principle of Creation, the branch of the Divine Principle book of Rev. Dr. Sun Myung Moon.

An asteroid is a celestial body, that is orbiting in the sky, made up of rocks and metals whose size can reach from a few centimeters to several kilometers.

According to the Principle of Creation, the interactions of particles have a continuous movement within each body and among bodies by creating the dynamism of all natural bodies in the universe.

Ryugu, discovered on May 10, 1999, is an asteroid measuring about one kilometer, which revolves around the Sun in an elliptical orbit, between 0.96 AU, at the closest, and 1.41 A.U., at the furthest (1 A.U. is approximately 149 million kilometers, this is the average distance from Earth to the Sun).

I believed that small natural objects in space were only animated by gyrating movements (translations) around a star. I was surprised when I read the news about the asteroid Ryugu. I found that this little star also had some rotating movements as we can read in the following expert :

"The asteroid RYUGU, measuring at least a kilometer wide in its largest dimension ... has a rotation period of about 7 hours and 30 minutes, this is the time it takes to make a

turn on it self's axis". (Cfr : futurascience.com, Hayabusa 2 : Mascot landed on Ryugu, by Remy De-court, published on 04/10/2018).

This is the detail from which cosmologists have drawn no conclusion as to the cause, but limiting themselves to observation.

Image by Ryugu

Image : JAXA, University of Tokyo, AIST

In principle, this rotary movement originated from the spins of the particles from which each natural body is resulted. The particles reflecting the dual characteristics of God move from the First Universal Energy contained within them.

2.2 The Principle of Creation and the Anthropic Principle

We are going on several occasions evoke the anthropic principle as being close to the position of believers with regard to the universe. We actually want to present those outlines.

The anthropic principle recognized scientifically, even if it is not unanimous, stipulates that the original physical parameters of the universe have been very finely tuned in order to make life and consciousness possible on at least one planet "which is none other than the earth" (Cfr : Wikipedia ; Multivers). If one of these pa-rams varied even slightly, life would no longer be possible.

The scientifically recognized anthropic principle, even if it is not unanimous, it stipulates that the original physical parameters of the universe were very finely arranged in order to make possible life and consciousness on at least one planet "which is none other than the earth" (Cfr :

Wikipedia; Multiverse). If one of these parameters varied even slightly, life would be no longer be possible.

Although this idea is considered to be philosophical, the anthropic principle has been used as an argument by Robert Dicke and Brandon Carter in cosmology and physics because, by doing so, the values of these parameters are exhausted. In this principle, man is the real reason of the universe.

This principle is confirmed by the Principle of Creation, chapter of the Divine Principle of Rev. Dr. Sun Myung MOON.

Why is it the Divine Principle of the Moon couple which confirms the anthropic principle and not its opposite? Quite simply, because the divine Principle is part of the cause, while in the anthropic Principle, we are only surprised to see the effects. So, we understand the reason why it is said that the Big Bang is confirmed by the Bible through the sentence "the separation of the water from the waters". (Ge 1 : 6)

However, the anthropic principle presents a quack that we cannot integrate into the Universal Cosmological Model, namely that it integrates the idea of parallel universes.

The Bible speaks of the pre-Big-bang, the Big-bang, the post-Big-bang as acts directed by divine will, and all in harmony. Everything in the universe has been positioned to allow life on earth according a model accepted by science called the Anthropic Principle. To arrive at created such a space, where things should only be shaped by collisions, in other words a physical phenomenon, until life can no longer be threatened in an extinctive way. Can we not thank such a God?

Let's take an example of a collision : according to science, the object called Theia struck the earth by injecting into space material that formed the moon. This collision, if there really had been one, tilted the earth's axis by 23.5; which causes the direction of the sun's rays to change throughout the year. If our planet had a perfectly vertical axis, these rays would always arrive at the same angle and there would not have been seasons. What a disaster for the survival of mankind. Was it not for this collision that cut the earth, what about gravity and our peaceful nights and days adjusting to our brains?

In addition, the presence of the moon or of the material which has been cut from the earth and which is part of our natural satellite to this day plays several important roles for humanity. Read :

Gen 1:14 God said, let there be lights in the expanse of heaven, to divide the day from the night; let them be signs to mark the eras, days and years; 15 and let them be lights in the expanse of heaven to give light to the earth. And so it was. 16 God made the two great lights, the greater light to rule over the day, and the smaller light to rule over the night; he also made the stars.

If Moses used the verb "to made" instead of talking about collisions as it actually happened, because the hammer of a star is perhaps just another star (meteorite), it is to demonstrate that it was part of a will. It's like on earth when you catch someone breaking something with a mortar. You cannot use the term accident or collision, but rather, break, separate or do something.

Conclusion on the principle of creation

The great question that we must ask ourselves is that of knowing why the study of the Principle of creation of Rev. Dr. S.M. Moon must be obligatory for all researchers as stated by Mrs. Hak Ja Han Moon in her numerous speeches?

First of all, no researcher feels satisfied with his or her intellectual limitations. Thus, an introductory quote from the Divine Principle says : "The Principle" brings about the unification of the inner truths that religion has explored and the outer truths that science has also sought". Thus, it is in the study of the Principe of Creation that one can discover the bridge that connects the spiritual and the physical.

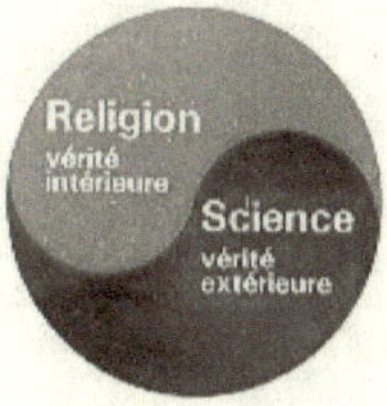

Source : The Divine Principle, level 4, p.13

Only through the study of the Principle of creation that we can remove the hidden veils from us and become free from all interior and exterior ignorance.

"The VERSETS OF THE SCHOLARS" Collection whose first part of the Volume I describes "The Universal Cosmological Model", is the first book to be able to restore scientifically the religious teachings and ancient philosophies dealing with cosmology. Note that it is thanks to the Principle of creation that served as the basis to build this Model. However, I leave the field opened to scientists for further research, and to write books themselves on the Physics of Creation, as well as some scientific versions of the Universal Cosmological Model.

Chapter V :
The Universe and the Cosmos

Section 1 : The meaning of the universe and the cosmos

1.1 The Meaning of the universe

Ge 1 : 2 "*The earth was formless and empty; there was darkness on the face of the deep, and the Spirit of God moved over the waters.* "

The universe is the name that keeps coming up throughout this work. So, before talking about the cosmos, we first want to explain this concept according to science.

In literature, the use of the term "universe" covers several meanings. Indeed, cosmology does not deal with the entire universe in general. Each concept evoked here deals only with a finite part of its totality. By the whole universe, we understand the universe as a whole. However, we do not know whether the latter is finite or infinite. In cosmology, we only take into account the visible universe which is called the observable universe, which is only a tiny part of the entire universe. We do not include in this notion what is still invisible, together with the visible one as we have said, form the entire universe.

The observable Universe is part of the visible Universe from the Earth through the light of the last objects that have reached us. It is often written with the capital letter "U". So far we have said that the most distant galaxy from the Earth discovered till now is called GN-z11. According to the calculation made by astrophysicists when they discovered it in March 2016, its light traveled 13.4 billion years before reaching the earth. Its actual distance is estimated about 32 billion light-years from the Earth. (Wikipedia : GN-z11).

The frontier of the observable Universe is called the "cosmological horizon". Currently, they said that it is located about over 46 billion light- years far away. It concerns a sphere centered on the Earth. It is the place where our vision through our most efficient engines allows us to see. But beyond that, it is only a deep darkness as the Bible confirms it as follows : *"Now the earth was formless and empty, darkness was over the surface of the deep"(Ge 1 : 1- 2).*

The universe regroups all the elements that surround us. While its contents are more or less known up to a certain horizon, its shape and total size remain mysterious or unknown to humans.

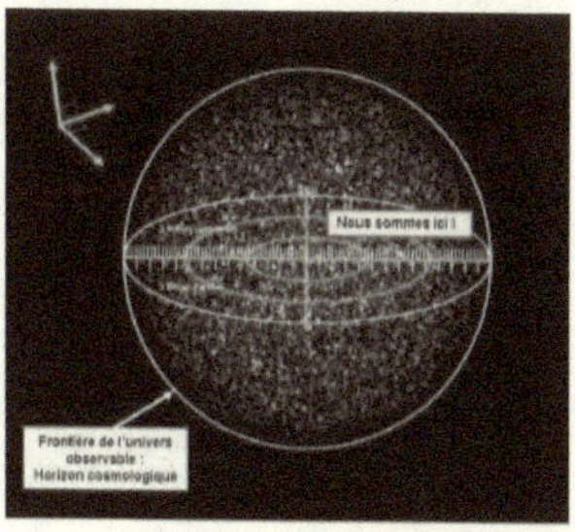

(Cfr : zevengeur.wordpress.com/2012/12/sphc3a8re-dunivers-observable- 03.jpg).

A photon (particle of light) that started its movement from this frontier in the past would have traveled at about a constant speed of 300,000 km / s, which is the speed of light, 14 billion by years, before reaching us. Consequently, from this result the astrophysicists took into account the expansion of the universe during its journey. It's like chasing a target which is still running from you. Thus, according to their calculations, the real distance undertaken by this photon would then have tripled in order to reach the earth. This distance, called the "comobile distance from the horizon," for this photon is going to be 14 billion years multiplied by 3 which results approximately 42 billion of light-years. (Cfr : https://zevengeur.wordpress.com/2012/12/sphc3a8re).

For each detected object at a long distance, a "comobile distance" can be calculated corresponding to the current theoretical position from the point of departure of its light.

1.2. The meaning of the Cosmos

Gen *1 : 1 "In the beginning God created the heavens and the earth".*

According to science, the cosmos represents the entire universe. Whereas the Rev. Dr. Sun Myung Moon, in his book about the Principle of creation, the branch of the Divine Principle, is the first scholarly religious to encroach on this realm, and behind whom we align ourselves, the cosmos is defined as the entire visible substantial world (entire universe) which can be compared to the human body; and the invisible substantial world (spiritual), which can be compared to the human spirit. (Cfr. Divine principle, level 4, p.35).

What difference is there between the scientific concepts and that of the Principle of creation of Rev. Dr. Sun Myung Moon regarding the cosmos? The Cosmos, as explained in the Principle of creation, branch of the divine Principle, encompasses all the concepts of the universe as explained by science, by integrating therein the spiritual world, accepted by religion, but excluded by science.

We know that the introduction of a spiritual sphere is added to the Principle of creation, even if some do not believe that; this is in the domain of the spirit and of the religion known as the spiritual world where the hell exists, the paradise and the Kingdom of Heaven, often referred to as the angelic world. Thus, the concept of the cosmos according to the Principle of creation is the branch of the Divine Principle of Rev. Dr. Moon responds to the Bible verse saying that "God created the heavens and the earth". The heavens represent the spiritual sphere and the earth the physical one. Here, it does not concern our planet, the earth in this verse. Biblically speaking, our present planet Earth was only created the third day (Ge 1:10). Here, we are talking about the land that was originally "formless and empty" which means the universe.

From the scientific point of view, we can call the Spiritual World by the concept that the scientists dislike, in the circumstances, "the subject universe" or "the cause universe". While according to the physical world, we can call it the "object universe" or "the universe of effects".

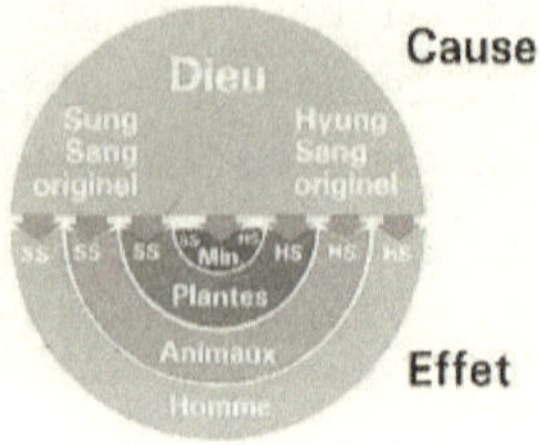

Source : The Divine Principle, level 4, p19.

2.1. **God's habitation**

Isa*40:22 "God dwells above the circle of the earth. "*

The top of the circle of the earth is what we call Heaven. So, it is very important for us to understand what this word represents.

2.2. **Heaven in the cosmos**

Gen 1:1 " In the beginning God created the heavens... "

Where is the Heaven and in which world planet is it located? Despite the abundance of religious literature, no one bothered to answer this question or explain it in writing. Hence, the importance of this book is for all who claim to be believers and especially the theologians.

Isa*66 : 1 Thus said the LORD : Heaven is my throne, and the earth my footstool. Where is the house you will build for me? Where will my resting place be?*

Believers say that God and the angels live in the Heaven. But if we ask them the question to know where this Heaven is, each will be content to lift his head. It is sheer ignorance.

By lifting our head so-called towards the Heaven, we are locating the geographic or physical Heaven which is different from the biblical language or from Biblical Heaven. Anyway, the geographic location of the Heaven also depends on the area where you are. Once you are in about a few thousand feet from the earth according to the aeronautical term, the geographic sky disappears and gives way to a Sky that covers you like a crystal bubble. Even to see the clouds while you are in a very higher distance in the air in an airplane, you look downwards.

In addition, when you are at a certain places of the universe, looking above at the sky, you are going to discover a dark space there. That way, we understand that the geographic sky is an abstract notion. The only appropriate term for Geographers to designate the Sky is the space.

Our machines fly either in the air or in space. The concept Heaven is purely religious. Science and literature borrow it without it responding to a real definition. What we see and what we call Blue Sky is only a limitation of our view or a panoramic reflection of the universe created by solar rays, passing through the atmosphere and other phenomena known to scientists.

The famous Russian astronaut Youri Gagarin said : *"I have traveled across the sky, but I have not seen God"*, confirming that God does not live in this sky that we see. From now on, Youri Gagarin is travelling across the real Heaven. Previously, he had only crisscrossed the space. And, in the air, you can see neither the angels nor God.

So, where is the true Heaven or the spiritual world which is also the holy seat of God? Let's read the following verses :

Ge 1.1 *"In the beginning when God created the universe." 2 Now the earth was formless and empty, darkness was over the surface of the deep, and the Spirit of God was moving over the waters.*

Ge *1 : 6 God commanded, "Let there be a vault between the waters to separate water from water" –and it was done. 7. So*

God made a vault, and it separated the water under it from the water above it. 8. And God named the vault the "sky". Evening passed and morning came – that was the second day.

In the above verse 2, we are told that before the Big Bang (the separation of the waters from the waters), the power of God which is the attribute of God, was moving in the water which was the representation of the original particles.

From verse 6 of Genesis, God parted this water ; this is the Big Bang : "*Let there be an expanse between the waters" and "God called the expanse Heaven".*

From now on, we have a clear definition of the Heaven. It is the expanse or empty space that separates everything and encompasses everything. This empty space begins even where I put my feet. This empty space which we know neither its beginning nor the end. This space which encompasses the entire observable and unobservable Universe contained in the cosmos.

We can say that geographic space is the visible part of the sky. Hence, it can also be called heaven, but, it is only content.

By saying that God is in the Heaven, it means that he lives in this empty space called the expanse by the Bible, the area which allows us to move like.

The galaxies and surrounds us. Hence the famous saying "God is everywhere". This is confirmed by the following words of Jesus Christ :

Mt 5:34 But now I tell you : do not swear an oath at all : either by heaven, for it is God's throne.

Mt 6 : 9 "This, then, is how you should pray : Our Father in heaven : May your holy name be honored".

Thus, in the upcoming project of establishing a city on the planet Mars or in other places in the universe we should never forget the settlement of a chapel for God ; who sees us in the invisible.

This is also the reason why the majority of people who pray prefer to close their eyes. It is a way of forgetting the material world or the distraction that the visible world can create for you, to find yourself immersed in that void which is Heaven, the habitat of God. And, we are all contained in this habitat. This is as we can read in this verse :

Jn 4:24 "God is Spirit, and those who worship him must worship him people worship him as he really is."

Sky view from the moon

"On the Moon, we see the earth bathing in the firmament under a dark sky. So, with the notion of the geographic and physical sky, a person living in the moon can never pray to "Our father who is in Heaven" without being able to explain to him the spiritual meaning of the word Heaven, because the physical context does not answer it". (Flavien PHANZU MWAKA)

We're not there also the atmospheric gases covering the earth, whether it be day or night, the sky will appear to us in this form. So, Biblical Heaven represents spiritual space encompassing physical space.

The most important aspect for God is not His Greatness or his omnipresence, but his heart. It is this heart that we seek to reach when we pray. So nothing is useful during prayer to lift your eyes or your head upwards and cry out aloud to be waited for by God.

Mt 6 : 6 "But when you pray, enter your room, close your door, and pray to your Father who is there in the secret place (empty space); and your Father, who sees in secret, will repay you "

Thus, after having searched the celestial sphere which is the habitat of God, but while residing spiritually in our hearts since Jesus Christ redeemed us, and physically in the empty space that surrounds us, He who is the source of all things, we now move on to the physical creation of the universe according to the scheme of Genesis of the Bible as restored in the Universal Cosmological Model.

2.3. The creation of the cosmos

The idea that the universe did have well and truly a beginning is known by scholars. It is even now stated into scientific explanations, including the famous "anthropic principle" according to which the original physical parameters of the universe were very finely tuned in order to make possible life and consciousness on at least one planet. (Cfr : Wikipedia ; Multiverse). For us, this is the work of God, but for the majority of scientists, it is "the coincidence of nature". But that still predicts that it had a beginning to get the parameters mathematically settled somewhere.

The anthropic principle in its scientific formulation is attributed to Brandon Carter, whose ideas are confirmed in the Principles of Creation, the branch of the Divine Principle of Rev. Dr. SUN MYUNG MOON, although he did not directly point to God clearly as being the true author of this universal regulation.

Gen 1 : 1 "In the beginning God created the cosmos (the heavens and the earth)".

HEAVEN : unquantifiable spiritual domain

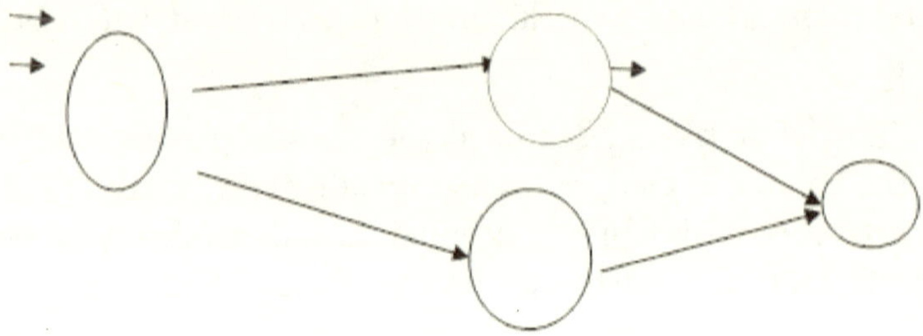

Earth (universe) : X-rays, cosmic ray, Galaxies : Quantifiable universe

Ge 1 : 2 The earth was formless and empty : something that exists, but we cannot touch it; therefore, formless, that is, the early Universe.

The cosmos, according to the Universal Cosmological Model, which is in line with Rev. Dr. Sun Myung Moon's Principle of Creation, represents the invisible (spiritual) world called the visible (Universe), both worlds are called "the earth" in the Bible. As it is previously written, the separation of two names in the diagram of the biblical creation by the conjunction (and), or even by a comma (,), means that both the heavens and earth were not created simultaneously. This separation, as we said above may take a thousands, if not millions, or billions of years.

Thus, when speaking of the creation of Heaven and Earth, the latter meaning the Universe, we see that the invisible world (or angelic world, or spirit world) preceded the creation of the Universe represented by the earth in the Bible.

The Divine Principle of the Moon couple defines the universe as the visible aspect of the invisible God.

The visible Universe contains galaxies in its empty space. Galaxies contain stars, planets, exoplanets, and other traveling objet such as asteroids, comets and meteorites. Planets have continents and water in various forms like the planet earth. The continents of the earth have countries.

Section 2 : Summarize

From this book, encyclopedias should modify their definition about the cosmos

– Cosmos

Ge 1 : 1 "In the beginning, God created the heavens and the earth"

- Spirit world (Heavens)
- Physical world (earth)

– Spiritual World heavens

- Kingdom of Heavens
- Paradise
- Hell

– Physical World

Ge 1 : 2 A formless and desolate land

Universe

– Universe

Space, total dark materials, dark energy, Galaxies

– Galaxies

- Stars
- Planets
- Exoplanets
- Other heaven bodies

– Planets and exoplanets

Chapter VI :
The Pre-Big-Bang in the Bible

For scientists, the front ”Big-bang” (or the front ”separation of the waters from the waters” by the Bible) is an area in which no theorist can venture, because, they say, the laws of physics no longer answer. In other words, the big bang front doesn't even exist, because time equals space. They are right insofar as for religious also it is a barrier that cannot be crossed. So it was part of the domain reserved for God or Christ.

It is true that after a careful scrutinization of the history of the biblical creation diagram (or the first cosmological history written on papyrus), we that this mysterious sphere, according to both science and religion, that we call the universe before the Big-bang, is related in the first day and, it was a very moving period. In other words, it was a period that did exist. Read :

Ge 1 : 2-3 "The earth was formless and empty : there was darkness on the face of the deep, and the spirit of God moved over the waters. God said : Let there be light ! And there was light... It was the first day".

Section 1 : The Basis of the Word Big-bang or Big Boom!!!

Before explaining to you the front Big-bang, we first want to explain what does it mean by the term Big-bang according to the referred image below :

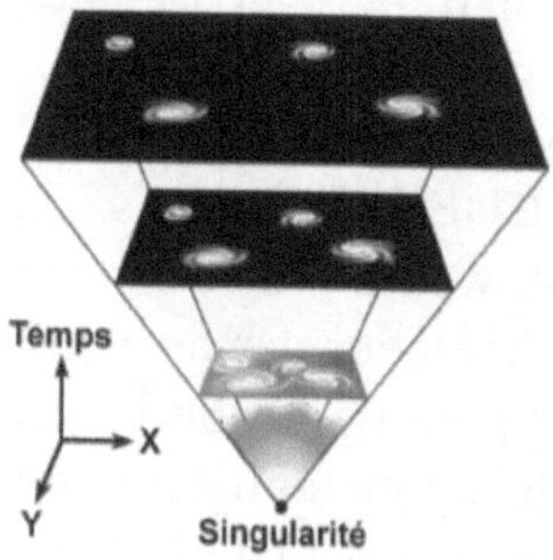

http://www.techno-science.net/?tab=glossaire&definition=

In this diagram, they are showing people that the universe at the beginning of its creation has only a ball form size as an atom, and which would have break out with a loud noise called in nowadays the Big-bang.

Some astrophysicists, according to them speak more about an expansion but not a burst. This starting point is considered to be a singularity. In mathematics, a singularity is a point where a mathematical object not well defined.

By making the Big Bang a singularity, scientists have created a barrier called "the wall of Plank" so as not to meet God. Yet the Big Bang took place on the second day of creation, according to the biblical story. If we did it without God, it would lead us to multiverse.

Based on various observations that we are going to describe in the chapters that follow, scientists state with almost certainty that the Universe was indeed born at time 0, which is called the Big-bang, or initial singularity!

This initial singularity is represented by a dimensionless "point" containing all the energy of the Universe. A point where all material and energy in the universe are condensed into a state of infinite density, along with all dimensions of space and time.

According to the Big Bang model, the present universe emerged from an extremely dense and hot state about 13 and a half billion years ago.

In the Science Documentary : "DID THE UNIVERSE EVER BEGIN? » ETIENNE KLEIN's lecture on "Science & You", the scientist, we said, distrust the singularity of the Big-bang which they consider as a transitional step, and they show that the study on the before Big-bang cannot be considered as being metaphysical, but physical instead. His lecture gave me a great comfort to present to you the following points with confidence.

1.1.1. The Pre-Big Bang or before Big-bang

The before Big-bang is the period which precedes this singularity, that is to say the Big-bang. For science, officially, this period does not exist and its study falls under metaphysics, that is to say a science beyond physics in the one hand. In the Universal Cosmological Model on the other hand, which makes the scientific restitution of the biblical schema on creation, this period did indeed exist and was characterized by real physical phenomena as we read

Ge 1 : 2-3 "The earth was formless and empty : there was darkness on the face of the deep, and the spirit of God moved over the waters. God said : Let there be light! And there was light... It was the first day".

In these verses, Moses describes the major steps of the state of the universe before the Big Bang. This great unquantifiable period ended with the appearance of the first light. Moses called it the first day; it has nothing to do with the time of our ordinary days. We have already explained the concept of biblical times in the introductory point.

Although all these sentences are contained in a few verses, the sequence of events shows that they are phenomena which do not take place at the same time. First there is a "shapeless and empty earth" meaning the Universe at the beginning where there was nothing inside. Then they tell us about the presence of darkness on the surface of the abyss. Finally, comes the spirit of God moving over the waters, which will result in the appearance of light. We understand that in this context, the word day simply means an unquantifiable period characterized by the accomplishment of some phenomena mentioned in the verses above.

Between an empty Universe and the appearance of water, the name used to describe the first original elements, we see that there has been a succession of physical phenomena.

This book scientifically unveils in a literal way, what this thesis hides for the consideration by science of the biblical diagram of creation, restored in this book under the name of the Universal Cosmological Model.

According to scholars, with regards to the complexity of the materials found almost everywhere in the Universe, they agree nowadays that the universe was initially made up of a mass which then exploded by emitting a great echo called the Big-bang. We have said that some scholars deny the existence of this echo, even though they agree with that phenomenon. What is important is the phenomenon itself. I have found that the same phenomenon is described biblically in the sentence "division of the waters by the water" (Ge 1 : 6).

Being the first researcher to find it out in the Bible, I made up my mind to scrutinize this concept in this book. No one expected the Big Bang to be Biblical or Biblically recognized one day. Some scientists even exhibited it as proof that the universe was not created by God. If the scholars of the 20th century knew that the Big Bang is biblical, they would have developed a lot of theories to contradict it. But at the present moment now, as far as they themselves have discovered the main proofs, the Universal Cosmological Model is only revealing and confirming it, and they can no longer go backward again.

Thus, we speak of the before Big-bang in science in order to designate the period preceding the appearance of the observable Universe ; and the post- Bang, the period which begins after the before Big Bang up to these days. These updates are made for my readers who have never opened a book of cosmology, likening it to atheist textbooks, being content with what is literally written in the Bible. It is not their fault, as long as we have said, the scholars were wielding the Big Bang to push God aside as the Creator.

Indeed, for scientists, the Big Bang constitutes a singularity, that is to say the beginning of all physically demonstrable things. But, for the Universal Cosmological Model, the Big Bang was a transitory stage, which puts God in his place. Why is this implied ?

Because this book is not a religion manual to confirm without proof that it was God who created the universe. This book only recreates in a scientific way the biblical diagram of creation, a diagram taken from the papyri, and from which everyone can draw their point of view on the Creator of the universe.

In my side or personal point of view, the scientific discoveries attesting to the veracity of the Bible, as emerged in this book have proved to me that it is God who is the Creator of the Cosmos; a point of view that I do not impose anyone to accept it. If anyone finds out from the results of my research that truly God is the Creator of the universe, then let only God be glorified.

We will discuss more, despite some intrusions of religion given my training and my status as a believer, the scientific phenomena of the original matter which led to the Big Bang.

It was the Russian-born American physicist George Gamow who first set up the physical theory of the Big Bang in the 1940s, which is articulated as follows :

"The Big Bang is the name of the official cosmological theory which attempts to explain the formation and evolution of the Universe. This theory assumes that all material (or energy) in the Universe has been found to be gathered at any given time; at that time, there was a phenomenal explosion which gave birth to the Universe"

(1) "Big bang, theory." Microsoft® Encarta® 2009 [DVD]. Microsoft Corporation, 2008.

Considering that this scientific theory of the Big-bang on the origin of the visible Universe is closer to the biblical diagram of the creation, in particular the second biblical day where one speaks of the "division of the waters from the water", we adopt this term in this book, for the harmonization of language with science. To be harmonized with the period as well, we consider the earlier Big-bang to be the period minus one of the universe creation period, with the Big-bang as starting from zero point. The Bible uses the sentence "divided the water," while science uses the word "Big-bang" which symbolizes the clash and noise that this separation made.

From the zero moment of the universe, it does not mean that the space- time has started there as most astrophysicists used to make people believe it today. The zero moment represents for us a new order of space-time of which zero constitutes the intermediate stage as confirmed by the French physicist Claude Sarrailh :

" *What you have to understand is that in the real world zero does not exist, it is a human invention from mathematics that determines a starting point. In reality zero is like all states, that is to say intermediate one«.* (Cfr : Claude Sarrailh, Le secret de la vie, éd. Librinova, Paris, 2018, P.124).

For the astrophysicist Michio Kaku in the scientific documentary "Beyond the Big-bang", the Big-bang is an error of designation, because there was neither Big regarding the condensation of the 'Universe that had become so small and no longer the "bang" sound, because there was no air to transmit the vibrations.

Indeed, to circulate, sound needs a support. It only moves through molecules capable of moving. These molecules are found in the air around us. Sound propagates by forming a wave in which each of the molecules energy encounters strikes another. However, in space, sound cannot be propagated nor be heard, because there are no molecules of air there. As the space is empty, sound cannot strike anything to propagate. (Cfr : P. Batchelor : Energy, sound and light, ed. New Horizons - ARS, Paris, 2016).

Through these clarifications, we see that the real terms are those that Moses used in the Bible, namely, "separation and expanse", which scholars have replaced with "expansion"; term used by the Belgian scholar Abbe Lemaitre who did not want his language to be biblical. And, this after P. HUBBLE confirmed that the galaxies were moving away from each other.

For science, it is the Big Bang that marks the beginning

of the observable universe, that is, the breaking out that led to the formation of the current Universe. This point agrees with what Moses said in the Bible in the following verse :

Gen 1 : 6-7 : God said, let there be an expanse between the waters, and let it divide the waters from the waters. And God made the firmament, and he divided the waters which were under the firmament from the waters which were above the firmament. And so it was. It was the second day.

These verses come to us, as we have said, from the papyri written around the year 1500 BC. AD and 3,500 years before contemporary science ; this is what all scientists forget and want to confront them with current scientific language. Instead of confronting them, we restore them to current scientific language.

Thus, the before Big Bang represents the period that existed before this biblical "separation of the waters from the water". Currently, scientists are unable to unravel the mystery hidden before the Big Bang, and they even refrain from advancing theories to avoid criticism. Others have tried to speak about, but they got confused in their own theory, because they put God aside. For me who is not a trained physicist, talking about the before Big Bang is tolerable.

Indeed, for scientists, before the publication of this book, the Big-bang was an absolute limit to the understanding of the beginning of the universe, because, beyond, the laws of physics are not valid any more nor even them, most basic concepts of space and time.

In other words, no man, according to scholars, has the right to speculate on the creation of the Universe before the Big Bang. The same point of view also applied to religion. Was required because for the moment this book has just broken the silence. So, we're going to see through this book, how Moses went beyond that reasoning. And, this book is the first to develop its mysteries for science and parables for religion.

Thus, with the Universal Cosmological Model which reveals the contents of great mass of the Universe before the Big bang, the scientists are going again to roll back time, by further studying the fossil radiation which would be the first radiation which appeared before the Big bang event. Big bang, described as "light" by Moses.

The particles of this fossil radiation, according to science has a particular importance to physicists, because most likely they are the key to future discoveries about the nature of material. (Cfr : Damian Grint, the mysteries of the universe, ed. Christophe Colomb, P.205, Paris, 1985).

The study of fossil radiation should also provide us with information on dark energy, dark matter and original matter. This thought goes hand in hand with the reconstitution of the history of the Universe after Big-bang by the scientists, who plan to do it with the radiation emitted by the stars as we can read :

"Reconstructing the history of the Universe is one of the primary goals of cosmologists and astrophysicists. The study of the dynamics of stars, and in particular their radiation is an essential step in the pursuit of this objective".
(Cfr :trustmyscience.com/telescope-fermi-retrace-histoire-rayonnement- stellaire-dans-univers/
by Thomas Boisson, 11/30/2018).

The Verses of the Scholars, by unveiling the biblical concept before the Big Bang, remove the stone that blocked the way for scholars. Glory be to God.

Section 2 : The first energies

2.1 The Universal Prime Energy

Ge 1 : 2 "The earth was formless and empty : there was darkness on the face of the deep, and the spirit of God moved over the waters. 3 God said, let there be light! light was. (Louis SEGOND). "

Ge 1 : 2 "Now the earth was empty and vague, darkness covered the deep, a wind of God swirled over the waters. 3 God said, Let there be light, and there was light (Jerusalem Bible)."

In these verses, within each sentence, there are energetic forces, even physical phenomena, which brought about the results.

Here is an illustration that inspires us to better explain the beginning of the universe :

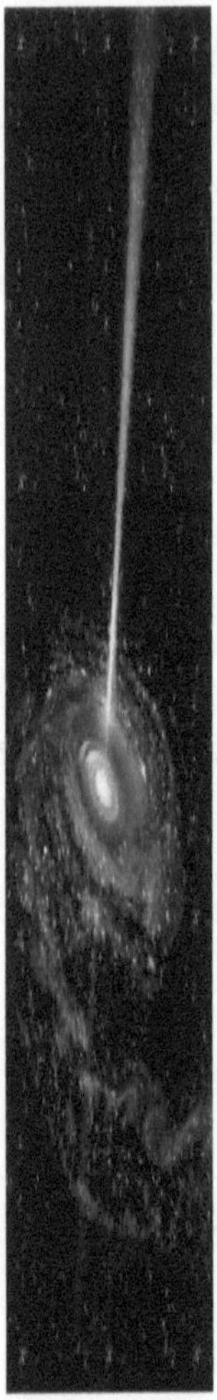

Artist's view of a rotating black hole. (Wikipedia)

For a good understanding by physicists of all that will be explained here in a literal
way, I recommend a scientific documentary confirming my comments on the above
verses. This is the documentary called "New Theory of the Unified Field :

Explanation of the Dynamic Universe"

In this documentary, everything came from a movement (a wind of God in the Jerusalem Bible's version). Regarding to the Principle of Creation, chaptered in the book of the Divine Principle of Rev. Dr. Sun Myung Moon, itis the movement from which all springs begin with Universal Prime Energy (E.P.U).

We can illustrate this through the following diagram :

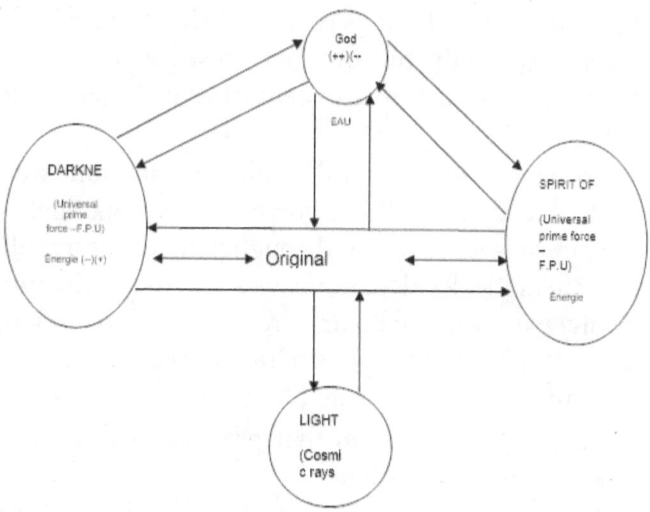

Ge 1 : 3 God said : Let there be light! And light appeared

Scientifically, this diagram reflects the gravitational contraction of the original particles referred to as waters in the Bible under the effect of very powerful energies called the spirit or wind of God. At the center of the pre- Big-bang universe, a space-time bubble is increasingly formed, described as an abyss which enlarges by accretion of the original energies, generates physical reactions by making light spring out. (rays) as it happens in the sun or the stars.

So while rays (light), whose zero charge would exert positive pressure, darkness which is the seat of dark energy, has been recognized by science as exerting negative pressure. We can read that in this quotation :

"Scientifically speaking, dark energy is the name given to a hypothetical component of our Universe, characterized by having a particular negative pressure"

What does the Bible say about the Genesis of the Universe, especially the prophet Moses who is credited with these writings many centuries ago before Jesus Christ?

When we follow carefully the explanation we are going to see that the Bible veracity is not parallel from what science is discovering through palpable observations and mathematical calculations. As for the formation of the Universe today, scholars are only making scientific discoveries about what has already been well written before, but using appropriate literary terms of that time.

Let us peel the situation that prevailed before the Big Bang according to the Bible.

Ge 1 : 2"Now the earth was formless and empty, darkness was over the surface of the deep, and the Spirit of God was moving over the waters. 3 God commanded, let there be light! And light appeared.

A shapeless and empty earth, meaning the universe at its beginning covered with darkness on its surface can be well understood by this phenomenon misinterpreted by scientists :

"One scenario suggests that before cosmic inflation, the Universe was cold and empty, and that the immense heat and amount of energy associated with the early ages of the

Big Bang was created during the phase change associated with the end of inflation. "(Cfr : Wikipedia : History of the Universe).

Indeed, the Universe was cold and empty, not before the Cosmic Inflation, but at its beginning when the biblical diagram of creation described it as shapeless and empty, and it took place even before the Big Bang. It began to heat up during deflation by condensing.

This is what is described through the Bible verse "*the spirit of God or a wind of God was moving on the water*" (Ge 1 : 2). Deflation also experienced some phase changes, as it occurred at a speed far greater than that of light. This speed decreased as the universe was created and condensed, causing the increase of heat caused by the shock of energetic particles during the collapse until the first appearance of light before the Big Bang.

The scientist Albert Einstein was the first to calculate this minimum speed when he reached the speed of light after the Big Bang, and he find the result of about 300,000 km /s.

The Verses "*Now the earth was formless and empty, darkness was over the surface of the deep, and the Spirit of God was moving over the waters.* ". *3 God commanded, let there be light! And light appeared* "(Ge 1 : 2-3) seem short, and yet it might fill a multi-page pre-Big-bang manual. These are short sentences for the readers, but concerning some very complicated situations that have taken place before the Big Bang. No scientist wanted to probe them. Its demonstration or scientific realization requires the capacity building of the Large Hadron Collider (LHC) in its current configuration as attested by this sentence below :

"*At the end of the 1990s, high energy physics seemed to be on the wake of an unquestionable revolution and perhaps even at the most important scientific revolution in all of human history since the physics of quantum gravity, that of the super-force dominating all others, and all types of material seemed to be reached in hand. We had to be disillusioned, because even if everything is still logically possible, it now seems much more likely that new physics is going to be ultimately accessible directly but only thanks to energies which are forever impossible to obtain on Earth".* (Cfr : futura science : Sterile neutrinos ; By Laurent Sacco, Published on 04/12/2017).

In any case, we can say that we cannot reject any hypothesis if we do not yet have the means to verify or fulfill it. However, we know that a claim only becomes scientific if it can be proven. Thus the idea of continental drift issued in 1912 was not accepted by all scientists until 1967. After all, in this 21st Century, science is still young compared to the age expected for the existence of humanity. We will return to this pre-Big-bang light in the following points.

2.1. The energy of emptiness

One of the questions that has always intrigued me in the first verse of the Bible is to know how from the spiritual one came to emptiness, and from emptiness to material. Knowing that according to believers, only God knows the answer, but, from a physics research perspective, this is part of the Phanzu paradox. In this book, I have mentioned the speed and collusion of rays several times without being a physicist. I had completely no idea one day that 10 years later I would find such an article on the web :

"When the right frequencies of light collide, they can turn into material thanks to Albert Einstein's famous $E = mc^2$ equation" explains Alberto Dominguez, astrophysicist at the Complutense University of Madrid" (Cfr : trustmyscience.com/telescope-fermi-retrace-histoire-

rayonnement-stellaire- dans-univers/
by Thomas Boisson, 11/30/2018).

So, in the future, when scientists will actually focus on the before Big- bang, the study of the evolution of heat from cosmic deflation will make it possible to quantify certain periods before the Big-bang. Although many puzzles remain to be elucidated, from the moment the word "earth" was firstly mentioned in the Bible, it was the beginning of physical processes into the Universe. Let us read it in the following extract :

Gen 1 : 1 "In the beginning God created the heavens and the earth".

All the scientific theories developed today, as having resulted from the formation of the primitive particles are phenomena which took place during the time of the chaos before the Big-bang according to the Universal Cosmological Model, and are now part of the physique of creation.

Indeed, the Big-bang being characterized by an exponential acceleration of the universe, called "inflation", it would have been impossible during this interval period of time estimated in milliseconds in this cosmic plasma that the elements are created starting from virtual particles to atoms within 10^{-43} (10 exponents -43) seconds to 3 seconds. It looks like the duration of an explosion; so, nothing could come together during this time. Rather, it was the time of liberation from what has already been created.

Here is an illustration of the Big Bang or the separation of the waters from the water :

Image :
https://qph.fs.quoracdn.net/mainqimg4d67c1855b85cf03
"With such violence during the Big Bang event, did the elements have time to merge in the space of a thousandth of a second to 3 seconds; starting from virtual particles, some cohesion particles - antiparticles through the strong interaction of quarks up to form the atom"? (Flavien Phanzu Mwaka).

What science has not been able to restore so far is that if the model of cosmological inflation has offered a solution to the problem of the horizon as well as to the problem of platitude (Cfr : Wikipedia; Cosmological principle), the scientists rather create some terms about the creation of the primitive elements starting from the virtual particles to the atoms. But this model has been presented until then as the one which also solved the problem of the constitution of atoms. If this is the case, scientists should also accept the biblical sentence saying that "God commanded and the thing was done" because even if you are a magician between "commanded and done" there is always a gap of seconds.

There is no way to accurately calculate the period from a nebula to a star, because a lot of things come into action like mass and gravitational force.

The formation of the concentrated mass of the Big-bang is even more complicated, because the material which composes it having been based on the collisions of rays of the virtual particles causing electric discharges which is unknown for the moment. In other words, it will continue until we discover the traces in the first lights (or radiations) that Moses speaks about. If it is the fossil radiation, this means that the studies, as we said above should still be done on these spectra.

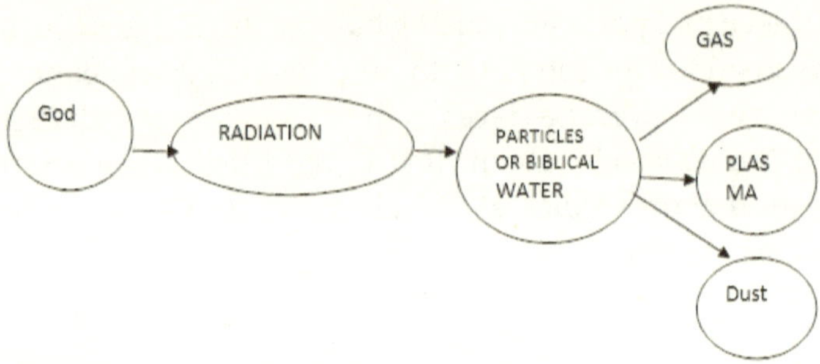

Rev 22:13 "I am the alpha and the omega, the first and the last, the beginning and the end".

This scheme can be understood through the following scientific sentence :

“Since the origins of the history of the Universe, the latter was created of a field of very high energy radiations which during a” hurly and burly” has generated material particles to the tune of about 1% from the initial energy. The latter is still present today with a very low energy density in the form of

the famous residual cosmological background ".(Cfr :
the famous residual cosmological background ".(Cfr :

zevengeur.files.wordpress.com/2018/07/instant-zéro).

All these reasoning of scientists, according to them concern the post-Big- bang period; but, with regards to the Universal Cosmological Model, the creation of the primitive particles took place during the before Big-bang period, even if after confinement, it firstly behaved like a plasma soup.

God is the source of energies. So, after having actuated the circular movement of the Universe, what Moses could have said by "God says that the earth rotates" ; everything began to be created from there. God is action. This truth, even science cannot contradict it. We can show it by this statement :

"The vacuum in quantum mechanics is not really empty. All the time, everywhere at each moment around us, pairs of particles and antiparticles are created out of nothing". (Aurelien Barrau, astrophysicist, interviewed by futura-Sciences, By Jonathan SARE, Futura-Sciences 08/01/2016).

Where did the energy in the chaos or the abyss come from before the Big Bang event? The theorists of the super-string theory speak about a three- dimensional star of the anterior universe comparing to ours now. As for the Universal Cosmological Model, we say that the universe being formless and empty at the beginning, so the chaos or the abyss was created from the original particles activated by the primary universal energy as it is written "the spirit of God was moving on the water".

Indeed, according to general relativity, energy has mass ; therefore, it gravitates. However, gravitation is attractive. (Cfr : https://www.futura-sciences.com/sciences/univers-energie-noire-4004)

According to the uncertainty principle, the collision of virtual particles creates particles when energy is supplied. The collision of these pairs of particles and antiparticles can only find their existence from what is developed here.

Until the appearance of this book, The LHC has not yet reached the phenomenal speed at which the Universe spins as we can read :

"You cannot reproduce in the laboratory the acceleration of free collisions or any other particle acceleration process.

We do not know how to reproduce the great distances and low densities of space, and phenomena cannot be properly miniaturized to laboratory dimensions". (Dr. David P. Stern Mail to Dr. Stern : education ["at" symbol] phy6.org).

Even particles so far subjected to the speed of light for their collisions up today will not be able to generate the original material referred to as "water" by Moses in the Bible.

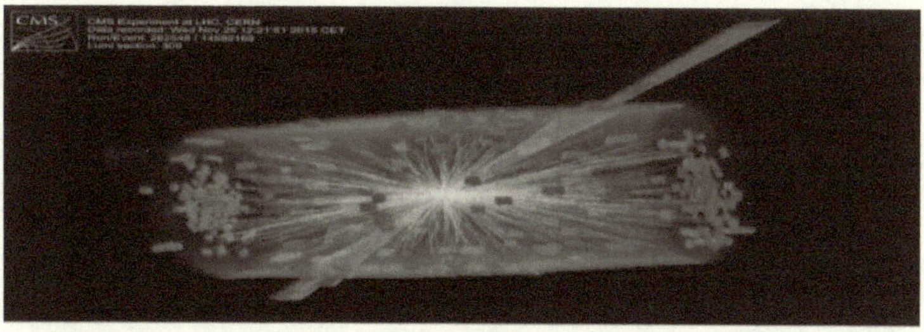

Lead ion collisions in the CMS detector (Image : CMS / CERN); to determine the state of material of the Universe at its origin

Science may be right when stating that the before Big Bang was a non- speculative period, as a Frenchman told me in a Blog. The latter said to me that :

"However, no one knows how to modeless this mysterious area with precision below the length and time of Planck, it remains an aporia". (Cfr : zevengeur.wordpress.com/2014/01/16/histoire - de - l'univers - Part 6 : Janus model, the future of cosmology ?).

The book "The Verses of the Scholars" updates the biblical energies that led to the Big Bang. Even science, in the quantum mechanics domain recognizes that at the beginning of the history of the cosmos all was energy. To speak about the creation of energies after the Big Bang as scholars try to speculate in the Standard Model does not stand. The problem is rather the place of quantum mechanics in the history of the Universe. Some scientists prefer to put it in the sphere of the Unknown Period of the Universe as we can see in this table :

Scientific representation of the unknown period

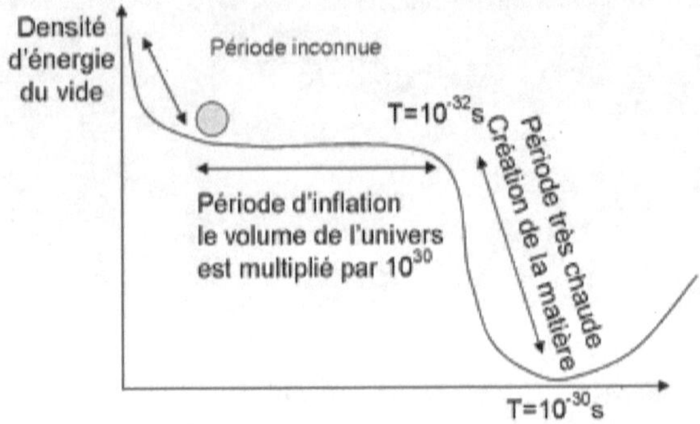

L'énergie du vide pourrait avoir évolué comme une bille roulant très doucement sur un plateau de haute altitude et en pente douce avant de chuter brusquement le long d'une corniche et de se stabiliser à sa valeur actuelle.

ces.org/articles/cosmologie-fastoche-33

Based on this table, several hypotheses have been made up concerning the uncertainty principle in order to explain the appearance of material after the Big Bang. This is the case with the following sentence :

"Even when the universe was created, while the material did not yet exist, the universe was agitated with a frantic succession of creations and disappearances of particles of all types. Moreover, it is thanks to this that ordinary material will be able to make its apparition«. (Cfr : annuaire-diggdes.coursgratuits.net/c7_bigbang/p715_partvirtuelles.php. By Olivier Esslinger).

For the Universal Cosmological Model, the principle of uncertainty - namely the quotation above should be deepened and positioned in the before Big-bang, that is to say, from the first universal energy up to ' the appearance of material. Everything that happened during the before Big Bang period was just a continuation of the deconfinement of what has been confined.

That is why, the further back the time of the universe, the more there will always be the question "and before that". Scientists say that any solution gives rise to new questions and this because the spiritual dimension of creation escapes us.

Section 3 : The great structures in the Biblical Pre-Big Bang

To help scientists in their quest, let us begin by explaining in a literal way even in a romantic one to my readers each structure about the origin of the universe before the Big Bang. This explanation is restitution in a scientific language of the diagram of creation according to the Bible. This biblical restitution is baptized under the name of the Universal Cosmological Model.

3.1. Shapeless and empty earth (universe)

Ge 1 : 2 "The earth was formless and empty"

To speak about the "universe" which is a specific term in cosmology, the common run of people prefers the term "world". Thus, a pastor who uses the term "world" to his audience or in front of the students will be better understood than if he were talking about the universe. This is how Moses had to use instead of the expression "primitive universe and without material existence" an explanation namely, "a formless and empty earth".

Let us only come back to the first word the "earth". This is different from the planet earth. Indeed, while speaking about an empty and unformed earth on the first day of creation, it would simply mean that the earth itself, as a visible material element had no existence. It is on the third day (in the 3rd step) of creation that the planet earth is discovered to have been materially formed. We can read it as following :

Ge 1:10 God called the dry land earth, and he called the cluster of the waters seas ...

13 so there was evening, and there was morning : it was the third day.

So what does the sentence, "*The earth was formless and empty*" (Ge 1 : 2) symbolize in the Bible on the first day? It is simply the universe at its origin. An empty and formless universe.

Mathematically speaking, Moses used the names of sets much more in comprehension rather than in extension, and in theology we speak about pictorial language based on forms rather than contents, because he wanted to demonstrate the existence of God the Creator to all people in general while writing these things, but not just only for intellectuals' class in particular.

Today, the scientific world uses many disciplines to explain the chronology of creation. But, in the time of Moses, who could understand these reasoning and these theories? Besides, even in the present day of scientific knowledge, the explanation of all this still hurts your head. Thus, Moses could not use words that people can easily understand.

So, when speaking of a formless and empty earth, Moses was only speaking of the Universe at its origin. Today, the word empty is debated in the scientific world, because it has been discovered that the vacuum physically is not necessarily empty, but it is filled with energy as attested by this article taken from the internet :

*"For quantum physicists, a vacuum is not anything : stripped of all material and at a temperature closer to an absolute zero, the empty space still contains energy! The proof of its existence? The "Casimir effect", capable of generating movement from this single energy «. (*Cfr : The force that comes from the void - Astrid Lambrecht in monthly n °

376/06/2004, p.48).

In the Casimir effect experience, it was discovered that two conductive plates placed very closer to each other spontaneously undergo an attractive force which can be measured.

This clarification is necessary to understand from what point Moses began to speak about the Genesis of the universe. In principle, it is from the following verse :

Ge 1 : 2 "The earth was formless and empty".

According to the quantum theory of fields and gravitation, the vacuum should have enormous energy.

(http://p1.storage.canalblog.com/18/78/910/17.jpg).

This theory is supported by mathematical formulas from Albert Einstein's general relativity.

Indeed, if the quantum vacuum does indeed contain a phenomenal amount of energy, nothing prevents that a part of it from being transformed into material, thanks to Einstein's equivalence relation (E = mc2). Thus, the origin of the universe would then have an immanent cause, because the void would be pre-existing the appearance of the universe. (Cfr : "Discourse on the Origin of the Universe", Etienne-Klein – 2010). All of this is confirmed by the Bible in the verse cited above.

What was there between the dark void and the energetic darkness ? The careful reading of this book and especially the writings of the Rev. Dr. Sun Myung Moon will allow scientists to draw deductions. Scientists will ask me why mix reason and faith ? It is simply because the universe was created by God and it is only those who approach him in humility who can draw great revelations like the pattern of creation or the Universal Cosmological Model. To men of science or reason to develop it. By saying created by God, we are not excluding the physical pattern. It is as if we are saying that such and such is the creator of such and such a device.

Nowadays, the observation that the Universe began in nothingness as described by the Bible to be a "formless and empty earth" is adopted by some scientists being atheists or not. Fred Hoyle's idea of a stationary universe did not have any support compared to a universe with a beginning. But it was the Bible that was the first book to talk about it many years ago at a time when no one could come up with such a conclusion, that is to say, at about 1500 BC.

According to the theory of the evolution of science, when man began to think about the origins of the cosmos, there was already the existence of the earth and the sky. Who could teach Moses that there was nothing at the beginning and chaos prevailed, too?

In view of the veracity of this statement about "a formless and empty earth" that means the Universe at its inception, and which is supported by some scientists, the Universal Cosmological Model now becomes the basic model in cosmology. As for the Standard Model, it has created a lot of question marks defined as riddles of the universe.

3.1.1. The notion of chaos and emptiness

Chaos and emptiness constitute, according to all religious, philosophical and metaphysical logic as the state from which the early universe arose. They are found under several synonyms such as nothing, nothingness, gulf and abyss.

Gen 1 : 1-2 "In the beginning God created the heavens and the earth. 2 Now the earth was formless and empty, darkness was over the surface of the deep, and the Spirit of God was moving over the waters. "

Surat al-Baquarah, 117 "Allah is the Creator of the heavens and the earth from nothing...".

3.1.1.1. The notion of chaos

"Chaos means, in the theory of the creation of the world established by the ancient Greeks, a dark and silent chasm that pre-existed all forms of life ... It should be noted that although some diagrams present Chaos as a sort of confused mass, all of them 'agree to see in it the first force to have invested the Universe" (1).

(1) "Chaos." Microsoft® Encarta® 2009 [DVD]. Microsoft Corporation, 2008).

Through this definition of the concept chaos, we rather fall back into the notion of the abyss which also means dark abyss and which is defined according to the glossary of Free bible as "*a mysterious world of the primitive ocean, place of darkness and evil powers, distress and perdition.* "

We understand that the notion of chaos refers to what happened in the biblical abyss, and nowadays in what astrophysicists have called a "black hole" or deformation of space-time.

3.1.1.2 The notion of vacuum

Ge 1 : 2 "The earth was formless and empty".

While talking about vacuum energy, we are referring to the energy over the early universe when the universe (the earth) was still formless and empty according to the Bible.

Indeed, the principle of scientific uncertainty implies that even the most absolute vacuum is in fact settled down with a myriad of particles and virtual antiparticles. Emptiness, as we usually understand it meaning the absence of all material as they do not exist as such. This is the ascertainment of scientists. (Cfr : astronomes.com/le-big-bang/particule-virtuelle/ August 9, 2017 by Olivier Esslinger)

According to science, vacuum is a state that preexists material and the movement that results from it. According to general relativity, the material- energy contained in the universe is not without induced effects, because it curves space-time in the latter. Moses called it an abyss : "there was darkness on the surface of the abyss." This is why quantum physics explicitly considers without referring to the before Big Bang together with space and time as pre- existing entities. In principle, it only confirms the Bible through its scientific restitution in the Universal Cosmological Model.

Ge 1 : 2-3 "Now the earth was formless and empty, darkness was over the surface of the deep, and the Spirit of God was moving over the waters..."

Wang Fuzhi (1619-1692) wrote : *"In reality, the principle resides in energy, and energy is nothing other than principle; energy resides in a vacuum, and a vacuum is nothing other than energy.* "(Cfr : JP LUMINET in Futurascience, the infinity of matter : the void, published on 01/25/2018).

In addition, Quantum Theory has shown that the vacuum was not really empty and that it contained a fabulous amount of energy, thus, the latter was then called "quantum vacuum". This void is filled with virtual particles that materialize briefly when energy is supplied, for example during high-speed particle collisions.

This "empty" state is indeed very different from a total absence. Hence, the term quantum vacuum used by science may therefore not belong to our universe according to the definition given to the latter. Hence this conclusion :

"The quantum vacuum, being filled with something could

according to this definition not belong to our universe. It could for example constitute the raw material from which our universe was created 13.5 billion years ago.(Cfr : zevengeur.wordpress.com/2014/02/11/The quantum vacuum : a new continent to explore).

This conclusion confirms what we have said, namely, in the Universal Cosmological Model that quantum mechanics with all the dances it offers us is not part of the post-Big Bang Universe. There is no place where you can insert these dances after the Big Bang characterized by inflation followed by an exponential expansion. Rather, it is a part of earlier Big Bang Universe, while according to the Bible "the earth was still formless and empty" meaning the early Universe.

When I came to this conclusion, I didn't know that I have either joined science or that the latter did the same as we can read it in the following extract :

*"Contemporary physics has not answered all the questions that it searched. Indeed, while it has made it possible to study many phenomena, the laws that it has highlighted do not meet in the areas where they are supposed to act jointly or concurrently. Essentially (but not only), quantum physics does not join relativistic physics". (*Cfr : matierevolution.fr, Dark matter, dark energy : the man-quant link ? 11/13/2007, by FaberSperber, Robert Paris).

Another strong argument is pointed out from Albert Einstein quotation, namely :

"For a billion particles of antimaterial, there was a billion and one particle of material. From a mutual annihilation, there is one billionth left : our present universe. "(Cfr : J. Baker : 50 keys to understanding astronomy, éd. Dunod, Paris, p.71).

This is what effectively confirms that the quantum mechanics, which according to us belong to the physics of creation should be inserted in the earlier Big-bang period.

Indeed, this quantum vacuum in quantum physics corresponds to an area of the Universe that supposed to lack material or rather in which nothing can be discerned. It is the area full of virtual particles and particles of antimaterial. When particles and antiparticles collide, they annihilate each other by releasing energy until they give rise to virtual particles in the quantum vacuum. All of these actions are described by the biblical verse The power of God was moving over the water".

All what we come to say above is well explained by the following scientific quotation :

"The energy of a vacuum is not zero, it is constant at all points in space- time which allows the law of conservation of energy to be respected throughout the universe. The energy of a vacuum is constant, but it fluctuates at this level and it is this instability of the vacuum that allows the creation of infinitely small materials. "(Cfr : Claude Sarrailh, Le secret de la vie, éd. Librinova, Paris, 2018, P.42).

As the earth (the universe) was initially formless and empty, the material as we see it today was formed from the energy contained in this vacuum, and whose properties come to us from the dual characteristics of God who has the power of bringing everything back to unity.

The Verses of the Savants are presented today as the suit that allows scholars to cross the arsenal of the Big-bang and still go up to a few billion years before the Big-bang. But, since there was no star of landmarks, this period cannot be quantified either by the Bible or by science. Moses, the scholar of scholars, had to contain this period which goes from the moment when there was only emptiness and chaos until the moment when the heat of the elements which caused the Big-bang reached a great heat, projecting the light, one day. Moses describes it as follows :

Ge 1 : 2-5 "Now the earth was formless and empty, darkness was over the surface of the deep, and the Spirit of God was moving over the waters.; and God commanded, let there be light! And light appeared... thus, evening passed, and morning came. That was the first day".

For these verses, we can say that there was a dark period in an empty Universe, filled with darkness on its surface, that is, energy. After having mentioned the title, namely "In the beginning God created the Heavens and the Earth", where many phenomena which are still inaccessible to us must have taken place, through the radiation by the universal first Force coming from the Universal Prime Energy, Moses began to speak of creation when the virtual particles began to move with the energies; and, that, by the phrase "The earth was formless and empty : there was darkness upon the face of the deep, and a wind of God moved above the waters" (Ge 1 : 2 - Bible of Jerusalem).

So, with an empty and shapeless earth understood as the original Universe, we move on to the second great biblical body of before Big-bang, namely darkness.

3.2 Darkness

Gen 1 : 2 "The earth was formless and empty : there was darkness on the face of the deep, and the spirit of God moved over the waters. 3 God said, Let there be light! And there was light. God saw that the light was good; and God separated the light from the darkness.4 God called the "light" day, and he called the "darkness" night. So there was evening, and there was morning : it was the first day".

In the above verses 1-4, we see that there was an energetically very eventful period before the emergence of the material world which scientists refer to as the earlier Big Bang.

As we said, Moses always considers God as the subject to show that creation was not a hazardous phenomenon as scientists believe. In the third verse (Ge 1.3), Moses speaks of the appearance of light. Now, if there is appearance of light, it means that there was darkness before (Ge 1 : 2) throughout the whole universe. This is also scientifically proven and this is what veiled these verses leading everyone to limit themselves to their literal interpretation.

If today all the stars went out, including the sun, we would return to a dark universe as in the beginning.

But, apart from 'God said', what is the scientific source of the light that appeared afterwards, especially since a light never appears on its own? By this we can say that the darkness here is not to be confused with our night darkness nor the light with our daylight. Darkness, in this biblical context, was made up of the energies that witnessed the creation of the observable universe. And some scholars even hypothesize their contribution, as we are going to read in the following paragraphs.

Here, we see the appearance in the darkness of the light which has no sources; That is to say, a light which has as its source neither the sun, nor the stars, nor any heavenly bodies, because at that time, neither the Bible nor science still speak of the creation of these things. The Bible simply says that God commanded, "Let there be light! And light appeared" (Ge 1 : 3). And God named this light "day" and this darkness "night".

Since "light" in this context does not represent our sunrays, so what about darkness? Since "darkness" is a state of something that we cannot touch, we say it should be a kind of energy that preceded the Big Bang moment. By asking science to know what kind of energies are involved in cosmology; we discover after much research that it should be dark energy, which is still the topic subject of many researches by scientists till today.

To those whom this name is quite new, we can simply say that the astrophysicists having observed certain phenomena that they cannot explain, including the acceleration of the universe, due to the distance between galaxy, have concluded that there is an unknown energy which contributes to this. They called it dark energy. Let us read this :

"Scientifically speaking, dark energy is the name given to a hypothetical component of our universe, characterized by having negative pressure. While the standard Big-bang theory, coupled with observations made so far predicted that the expansion of the universe would slow down under the effect of gravity, it seems to have accelerated over the past few billion years. This is how the idea of an invisible energy came into being, which has been renamed dark energy and which today has been scientifically proven since 1998 by two teams of astrophysicists led respectively by Saul Perlmutter and Adam Riess "(Cfr : 13/11/2015 -Laurent Sacco : Futura Science on web).

As we have said, Moses had already discovered it and has called it darkness. An ancient philosopher whose name is Empedocles also spoke about this energy and nicknamed it hatred which is responsible for discord or acceleration of the universe. It is a great first for astrophysicists to hear that dark energy was already described in antiquity. This is to say that it is not a word matching game. It was wrong to think that all the scientists of antiquity could not reason like our tri-docs PhD in natural science today.

Indeed, the scientific sentence describing dark energy as it is done both in Wikipedia and the Bible is :

Bible : *« The earth was formless and empty; darkness was over the surface of the deep..." Ge 1 : 2.*

Science : *"In cosmology, dark energy is a hypothetical form*

of energy uniformly filling the entire Universe" (1)
(1) Dark Energy : wikipedia.org.

In the scientific documentary "New Theory of the Unified Field (Explanation of Gravitation)" that I discovered in 2018, without referring to the Bible the author may never have read, demonstrates that everything began with the dark energy that filled the universe, and the movement on these photons was the source of the creation of some elements. Religiously speaking, this thought of the author describes the following biblical sentence : *«The earth was formless and empty. darkness was over the surface of the deep, and the Spirit of God was moving over the waters*" (Ge 1 : 2).

Whoever says "wind of God" according to the Bible (Jerusalem version) sees the movement notion in science; and the one who says "water" according to the Bible means the "elements" according to science. Note that in ancient times, the language of science was not as developed as it is today.

In the same context, the physicist Claude Sarrailh said :

*"...the movement is Creator of material, and the vacuum is a ceaseless movement of virtual particles which exchange with the particles-material or those which are not yet created». (*Cfr : Claude Sarrailh, Le secret de la vie, éd. Librinova, Paris, 2018, P.109).

Science assigns several names to this primordial energy, namely, dark energy, cloudy energy, void energy, etc. This primordial energy contributes to the generation of dark material according to the documentary cited above, that is to say, which is everything that cannot be seen in the Universe. Everything that is still invisible to us until today. We can estimate it about 95% throughout the Universe, and its existence is beginning to be attested by science nowadays.

So, such energy in our opinion, which controls the expansion of the universe after the Big Bang, cannot appear after it, as the scholars say. In addition, according to scientists, it controls our destiny, as another scientific analysis again asserts in this quotation :

"Dark energy may also be the residual trace of an inflationary period in the very early history of the universe. If so, then it is deeply linked to the very apparition of the observable Universe, the material it contains and even the formation of galaxies and galaxy clusters. What is almost certain is that it has to control our destiny. "(Cfr : Laurent SACCO, www.futura science, 04/10/2011).

Those are statements which join the Bible and which support the Universal Cosmological Model.

Biblically speaking, given that darkness (dark energy : negative according to science) and the power from God (positive energy) were in action before the Big Bang, it could be very difficult for us to understand their history and destiny. Even Moses does not mention how the forces that gave rise to these two energies began. For the Rev. Dr. Sun Myung Moon, with whom we agree, these two energies come from God who is the source of all energies. (Cfr : The Principle of Creation).

Currently dark energy is recognized by science as a negative force which provides the expansion of the universe. There remains an enigma for the new science today which is unable neither to detect it nor explain its origin. But, as in the Standard Model, the Big Bang is considered to be the starting point of the origin of all things while scholars say that it was created during this rupture.

In the Bible, on the other hand, darkness is the seat of this dark energy, and it did exist well and truly long ago before the "separation of the water from the waters" or

Big-bang. Under divine inspiration, Moses called it "night" even before the creation of the sun and stars. It means that the word, "night" in the Biblical scheme of creation has nothing to do with our present nights caused by a change in the weather.

If it is also necessary to come back to dark matter, which I say is also a vestige of the pre-Big-bang era, while the manuscripts of this Volume were being revised, a scientific article was published going in the same direction as my research. This was done by the astronomer and physicist Tommi Tenkanen of the Johns Hopkins University, who we can read :

"If dark material was really a vestige of the Big Bang; in many cases, the researchers have already seen a direct signal from dark material in different particle physics experiments". (Cfr :trustmyscience.com/nouvelle-etude-matiere-noire-plus-ancienne-que-big-bang/. By Stéphanie Schmidt, August 9, 2019.)

For us, this scientific article also confirms the Universal Cosmological Model and now calls researchers to think about it.

3.3. The abyss

Gen 1 : 2 "the earth was formless and empty, darkness was over the surface of the deep, and the Spirit of God was moving over the waters. 3 God commanded, let there be light! And light appeared ".

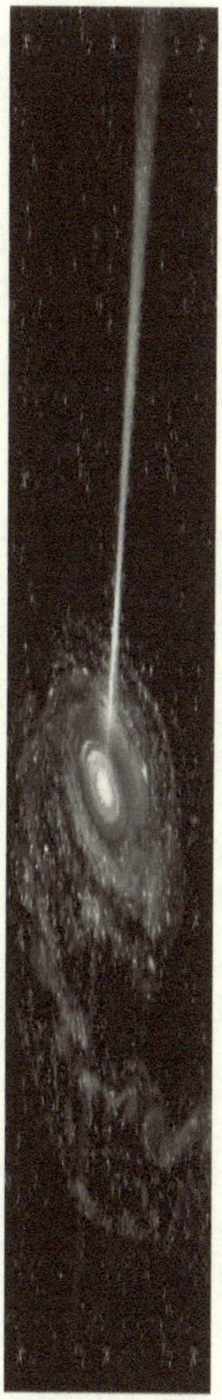

Artistic view of a rotating black hole (Wikipedia).

In the biblical glossary taken from Free Bible, the abyss is defined as "a mysterious world of the primitive ocean, a place of darkness and evil powers, of distress and perdition". Thanks to this definition that I came up with the idea of referring to black holes in the Universe.

Even scientifically speaking, this definition does indeed

agree with the explanations of what scientists call the black hole. Who knows the contents? The primitive ocean effectively represents the cluster or emptiness in the beginning of the Universe, a place of darkness and energies in confrontation, a place of no return (distress and perdition).

Indeed, before undertaking the realization of this work, I always used to read this verse in the Bible, and like all the common run of people, the word abyss meant to me like a hole or an empty space in a closed place. But, why didn't Moses directly speak of a hole or empty space? Quite simply, according to the above definition taken from the Free Bible glossary and, of which Moses was not aware about it, we can say that the abyss is a hole in which mysterious phenomena take place. The glossary even adds the word perdition, that is, once caught there, there is no longer any hope of leaving. the only things that agreed with the above explanations that astrophysicists have discovered is that the abyss is like black holes in the sky.

For a small explanation, a black hole is an empty space in the airspace of the sky endowed with a gravitational field so powerful that nothing can come out of it once its "horizon" (or point of no return) is already achieved. Scientists themselves do not give a clear definition concerning it. They prefer only to describe the phenomena taking place there.

The concept abyss used by Moses in the Bible literally fits with what physicist John Wheeler scientifically has explained about black holes as we can read it in the following expert :

"The term 'black hole' was invented by the American physicist John Wheeler in 1967 to describe a concentration of mass-energy which gravitationally collapsed under its own

force of gravity and which has become so compact that even photons cannot escape this gravitational force"(Cfr : Laurent SACCO, www.futurascience, Trou noir).

For Americans, the explanation below is tolerable : the black hole is like a tornado in interplanetary space which also generates a large whirlwind, except that one cannot see the bottom of it. In the Earth Tornado, things are inspired from the lower base and propelled upward. While in a black hole, materials are sucked in from the sky (resembling the eye of the earth's tornado; but, for black holes, we speak more of the mouth). The whirlwind of the black hole is so violent so that it is even capable of attracting and swallowing in one mouthful a planet larger than earth; mash it forever, or inject it as debris and dust into space. The luminous jets would precede this injection.

Nothing can escape radius's gravitation of the black hole which is called the event horizon. Everything that crosses it, including light plunges into it. Hence its definition in biblical glossaries as being *"places of darkness and evil powers, of distress and of perdition."* "

If scholars did not distort the biblical language by discovering black holes, John Wheeler might have called it the abyss (1967). But as we see, scholars always used to bypass the biblical language in order to give inappropriate names. Till today, they continue to say that man is made from the dust of the stars in order to distort what the Bible says, namely, "man is made with soil taken from ground of the earth" (Ge 2 : 7) Now, whether man is made of star dust (upstream), he was formed on earth (downstream).

When I started the realization of this work, the word abyss troubled me a lot, and I thought in myself that it could hide the secret of the creation of the universe. By associating it with black holes, more troubles concerns heightened again, because according to current science, the holes appeared after the Big Bang. And later on, when I saw in the images that the distortion of space-time is presented as an abyss, I was happy. Scientists could not escape to this matter which is mine also. Let us read this sentence quoted from futura science :

"It is easy to understand why physicists are so excited to study black holes, because these objects are probably a key to all fundamental physics and to a deep understanding of the origin of the Universe."(Cfr : futura-sciences / definition / astronomy-black hole).

In other words, through everything that is explained in this chapter, the Universal Cosmological Model contributes greatly to science.

In the scientific documentary about "The Mysteries of the Universe", it is said that black holes keep galactic gases, this is how they attract all surrounding gases to themselves. Its attraction, called gravitation by scientists it is so powerful that even if it engulfs a star like the sun, its surface is going to remain dark. Sometimes it can also throw it up as we said after pounding it, and that by a jet of great light.

It is for the first time that non-cosmological believers learn that whirlwind- like phenomena also take place in the profound sky.

For my many readers who have no idea about the tornado, especially Africans, I offer here an illustration for better understanding of the black hole into space :

(The base)

(TERRESTRIAL TORNADO OR WHIRLWIND CREATED BY THE WIND : the base and the top

An illustration of what can be the celestial black hole, described as an abyss by Moses or the first zone of concentration of material (Ge 1 : 2). [Universal Cosmological Model.].

We recognize a tornado when we spot a whirling, spinning funnel falling from a large cloud. But, in a black hole, we cannot see the lower part. Just as strong swirling winds frequently emit a shrill hissing or whistling sound that can be heard from several kilometers away; it is the same for holes with some sounds which can be picked up by astronomical devices as we can read it in this article :

"Researchers have finally succeeded in penetrating the mystery of the radio signals coming from the black hole in

the Arp 220 galaxy". (Cfr : fr.sputniknews.com : a black hole emits mysterious sounds).

In space, the whirlwind of the black hole looks like this :

A supermassive black hole stripped of its galaxypassed through the universe like a whirlwind in complete freedom and at high speed by Stéphanie Schmidt,November4,2016,trustmyscience.com. (Cfr:https://trustmyscience.com/un-trou-noir-supermassif)

Schematically speaking, the force around the black hole looks like this :

They are usually represented as gigantic siphons capable of distorting the space-time around them (the blue lines in the illustration). (Cfr : Joe Bergeron, Encarta ® 2009. Modeling a black hole).

These images do indeed meet the definition of Moses who said, "*...darkness was over the surface of the deep, and the Spirit of God was moving over the waters. God commanded : let there be light! And light appeared*". (Ge 1 : 2-3).

The meeting of all the lines forms the bottom of the funnel in the above picture defined as the abyss by Moses in the Bible and existed when the Universe was empty. The work movement of the spirit of God or a wind of God is represented by the lines. At the bottom, we discover what astrophysicists call the wormhole. Above, it is the light which springs from the fusion of the original material (of the water) and which illuminates the space (God commanded : let there be light! And there was light appeared). Hence the biblical sentence : "*when he established the clouds above and fixed securely the fountains of the deep.*" (Prov 8:28)

According to science, the Universe recognized itself condensed before the Big Bang. But as it condensed, the material that it confined was called "waters" by the Bible as a result of its condition. Indeed, before having the Big-bang or separation of the water from the waters according to the Bible, we need to be around a zone of attraction where the original particles began to intersect according to the literary term, creating shocks by generating heat and light which finally produces the Big-bang. It is this zone that Moses called the abyss, and that what we call the first black hole in the universe, or the first deformation of the fabric of space-time.

This fits with the following scientific description :

"In the framework of Einstein's equations, the beginning of the universe must correspond to a very chaotic and turbulent state of space-time which could force its content in material or in light to collapse locally by giving black holes of masses which could be much smaller than a planet or even of an atom". (Cfr : Laurent Sacco, www.futura science, the death of Stephen Hawking leaves us alone in front of the enigma of the Big-bang, published on 03/15/2018).

In addition, the uniform temperature of the cosmological diffuse background reveals that the universe had indeed been condensed, which allowed all the extremes to benefit from the same temperature rise. The spirit of God according to the biblical terms has created what we called the abyss, or an area of attraction of material named the universe before the Big Bang.

Moreover, the idea of science according to which the Universe emerged from the black hole (abyss) begins also to gain the ground as we can read it in the following scientific commentary appeared in Futura-science.fr :

"An ordinary black hole is created by the gravitational collapse of a massive star; it is a local event in the first hand. Big bang models, on the other hand are events on a global scale; so it is difficult to relate them. Nevertheless, we start to consider models in which there is a front Big-bang. We can imagine and calculate solutions which put in scene of a Universe which was in generalized contraction for example. This Universe, instead of collapsing on itself in a Big Crunch it would have formed a black hole. However, according to this theory we know that under certain conditions the "bottom" of a black hole is not necessarily blocked : it can produce a wormhole. This wormhole would be a kind of tunnel that could connect the entrance of a black hole to an exit : a white hole or a white fountain. It's quite tempting to imagine that our Big Bang could ultimately be the outcome out of a wormhole. This is one of the few hypotheses on which researchers are currently working". - (Cfr : Jonathan SARE, www.futura-Sciences, published on 05/11/2014).

Another argument is taken from the science documentary entitled "The End of the Universe" in which Professor Kimberly WEAVER said that *"the physics of black holes is closer to what gave birth to the Universe."*

Once again, this reasoning assumes the probable existence of a black hole before the Big Bang event. This rejoins the biblical sentence about the front Big Bang which says that "*The earth was formless and empty.*" Later on, through the action of a wind coming from God, according to Jerusalem version of the Bible, an abyss was formed as it summarized by the following biblical sentence adapted to the understanding of the antiquity people : "*The earth was formless and empty, there was darkness on the face of the deep* "(Ge 1 : 2).

Through this, we see that science is moving without knowing it towards what Moses has said, then it is towards the Bible.

The concept abyss used in the Bible, which constitutes the great key to better understanding for the beginning of the Universe, represents a kind of black hole created by the "condensed Universe". This is the starting point from where everything was contracted before going through a rebound experience. And this concentration of mass creates according to Albert Einstein "the deformation of the fabric of space-time" which has almost become a name reserved by science to explain the shape that the fabric of the universe takes in contact with the great masses.

Could it be a word game between the terminologies "abyss and black hole»? Not at all, insofar as the black hole illustration does indeed respond to what Moses has described, as we can see it in the photo below :

"Hidden behind thick clouds of absorbing dust, the closest supermassive black hole to Earth is located at about some 26,000in the light years away, inthe center of the lacteal Way (see circle)." (Source : https://www.techno-science.net/illustrat/Source/ESO/2018/07/26/eso1825a.jpg).

In this illustration, we don't see any hole except the darkness with stars. On the other hand, the astrophysicists noticed on the basis of observations carried out by means of the Very Large Telescope of ESO in Chile, a small group of stars orbiting at a higher speed around the circle, and they concluded that there exists a super solid black hole. This effectively reflects earlier biblical Big-bang sentence which says that "The earth was formless and empty, there was darkness upon the surface of the deep" (Ge 1 : 2).

Despite of the darkness, Moses knew how to describe the existence of an abyss at the beginning of the universe in the same way as the black holes discovered today by science. He even describes its reigning gravitational force not only around, but also inside it as well as in the following sentence *«the Spirit of God was moving over the waters."* (Ge 1 : 2). The water represents in this illustration thick clouds of absorbent dust. This gravitational force is exposed through the motions of stars orbiting. (Cfr : www.techno-science.net ; successful test of Einstein's theory of relativity near a supermassive black hole by Adrien "07/27/2018).

The original "black hole «as designated the abyss by Moses is now reduced to thousands of black holes by the clusters it has created after the Big Bang, and that scientists call "micro - black holes". This assertion was supported by the assumptions contained in the scientific documentary entitled "Unexplained Phenomena of the Universe" written by Laura Verklan Armstrong.

It is said in this documentary that "the mini black holes could be the vestiges of the Big bang". Science uses the verb "power" (for lack of a better term) without any scientific evidence. The same ideas are made by NEIL TYSON from the US Museum of Natural History as follows : "black holes are perhaps the embryos which allowed galaxies to be formed" in the scientific documentary entitled "The Milky Way and its huge black hole", written by the director Thomas LUCAS.

For us, the existence of black holes is also due to the exponential expansion of the universe, demonstrating that the evolution of material was lower than the rates of expansion. But, this finding has nothing to do with the word imbalance. This is what NEIL TYSON wanted to explain in the rest of his remarks on the same documentary by speaking of the "mixing of material after the Big-bang".

I know that the explanation of the abyss has spilled much ink among scholars. Once the Universal Cosmological Model is adopted, this word risks creating two schools. That of the Universal Cosmological Model with black hole and that of the Universal Cosmological Model without black hole, that is to say, only with a deformation of the fabric of space-time.

3.3.1. The biblical abyss as a black hole

For the first school, the abyss would be assimilated to the black hole before the Big Bang from which the observable universe would have emerged after having been prepared in this gear. Several scholars who attempt to think about the before Big Bang adopt this hypothesis.

Indeed, in the black hole, there is a gravitational force which even light cannot escape. Hence the biblical phrase : "there was darkness (taken in the sense of darkness) on the face of the abyss" [Ge 1 : 2]. This is what is seen on the surface of a cosmic black hole.

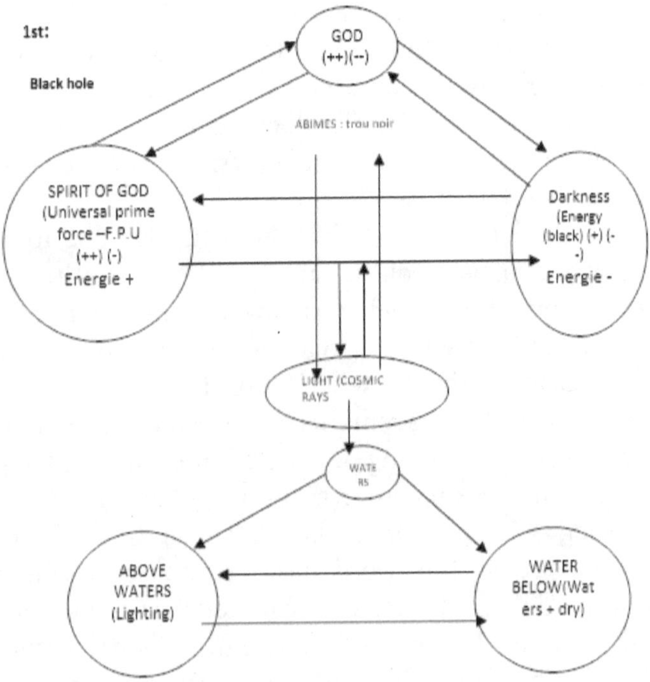

(Ge 1 : 2-3... darkness was over the surface of the deep, and the Spirit of God moved over the waters. God said, Let there be light ! And there was light).

What other scientific evidence in the observable universe that can show that things originated from black holes ?

– On the one hand, there is an overpopulation of young stars near the galactic centers. Simulations have shown how they can be produced and ejected by the accretion disks of supermassive black holes.

– On the other hand, for some scientists, the main mechanism of galaxy formation is the collapse of a gas cloud, perhaps around the primordial Black Holes.

3.3.2. The biblical abyss as a distortion of the fabric of space-time

For the second school, with regards to the beginning Moses defined the Universe as being empty. From the moment when the first elements started to agglomerate, it changes status and goes from infinite void to an infinite empty hole, but containing something ; that is, thus becoming an abyss.

For example, in an empty Universe, we cannot speak of a hole. But the moment there is energy that begins to condense in this void, Thus the Universe becomes a hole called an abyss or a deformation of the fabric of space-time according to the scientific term. So the notion of the deformation of the fabric of space-time attributed to Albert Einstein is rather biblical. Moses who was the first to describe the phenomenon used the word abyss. It does tie in with science.

Speaking of materials, we can already see a long way. According to engineers Ruffin MPANZU KIADI BABAKA and NIMI MALONDA :

"For Albert Einstein in general relativity, any source of energy already causes a local deformation of space-time; the curvature of space-time is therefore the manifestation of gravitation, more precisely a manifestation of the gravitational field of the source which is at the origin " (Cfr : Ruffin MPANZU K.B. and NIMI M., National Pedagogical University [DRC] ; explanation of quantum gravity, T.P. 2017, unpublished).

Example of a distortion of the fabric of space-time or damage according to the Universal Cosmological Model

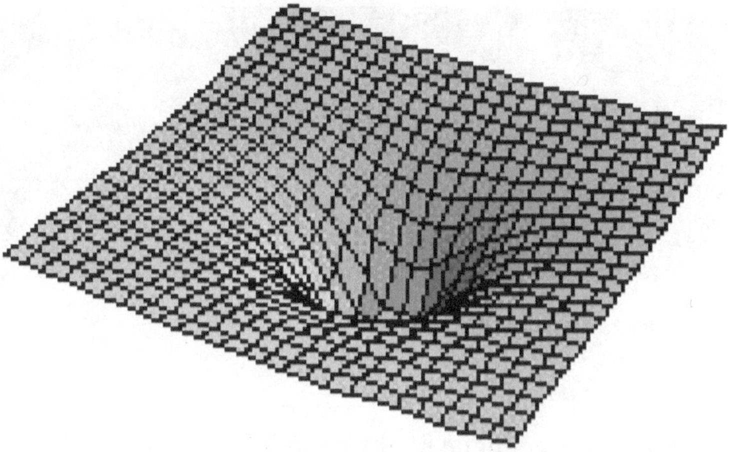

(Ge 1 : 2-3 ... there was darkness on the face of the deep, and the Spirit of God moved over the waters. God said, let there be light! And there was light).

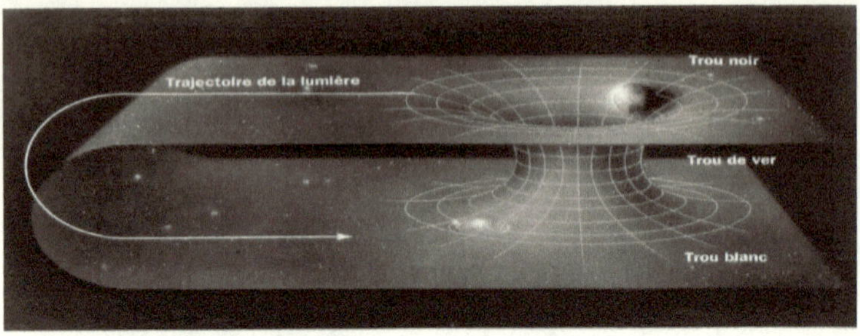

Schematic explanation

http://maclasseenligne.fr/physique/images/trou/trou%20noir

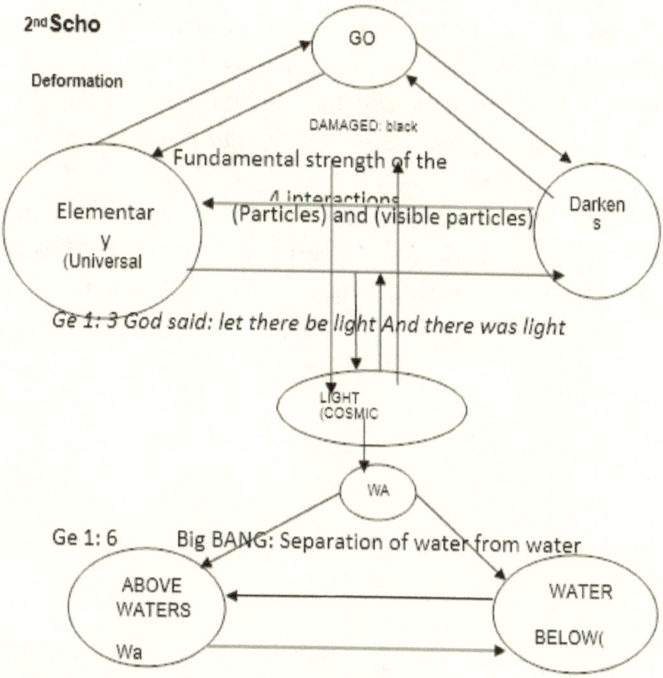

Let's read this verse again :

Ge 1 : 2 "The earth was formless and empty; there was darkness at the face of the deep."

What is the position of the Universal Cosmological Model with regard to the two schools on the word abyss? We leave the appreciation to science, of which the two visions are already a part. Indeed, general relativity describes a dynamic space-time, in interaction with material and energy, while in quantum physics, space-time is only a passive framework where several phenomena take place. These two visions seem contradictory and must be reconciled.

Nevertheless, we confirm the existence before the Big Bang, whose original materials were concentrated, as the following verses further attest :

Pr 8 : 26-28 "before he made the world or its fields, or any of the dust of the earth. I was there when he set the heavens in place, when he marked out the horizon on the face of the deep, when he established the clouds above, and fixed securely the fountains of the deep,"

These verses, from the same divine source, again summarize Ge 1 : 1-10 in detail and confirm that the early energies, by concentrating, dug the abyss into the fabric of space-time.

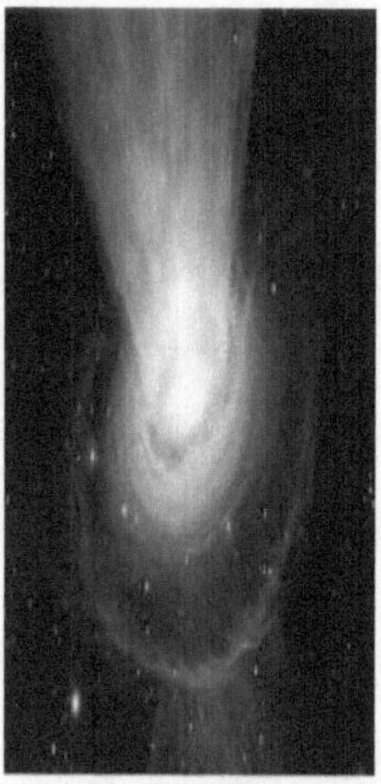

https://cdnfr2.img.sputniknews.com/images/103165/96/1031659637.jpg
Pr 8:28 "And the sources of the abyss gushed out with force" or Big-bang.

BIBLICAL MODEL

VISION OF KING SOLOMON		WRITINGS OF MOSES		
Pr 8 :26	he had not yet made the earth (COSMOS)...	Ge 1 :2	The earth was formless and Empty	
Pr 8 :27	when he laid out the heavens	Ge 1 :1	God created the heavens	
Pr 8 :27	when he drew a circle on the surface of the abyss	Ge 1 :2	... There was darkness on the surface of the abyss	
Pr 8 :28	when he fixed the clouds above	Ge 1 :2	the spirit of God was moving above the waters	
Pr 8 :28	And the fountains of the deep gushed out	Ge 1 :3,7	He divided the waters which were below the firmament from the waters which were above the firmament: Big –bang	
Pr 8 :29	When he laid the foundations of the earth	Ge 1 :6-10	God called the dry earth	

Who was Solomon to whom the writings of Proverb are attributed? He was the son of King David and succeeded him to the throne. He asked God for intelligence to rule Israel, and God promised him that there will be no one equal to him in matters of intelligence and wisdom.

Read :

1R3 : 11 "And God said to Solomon, since this is what you ask, since you ask for

yourself neither long life, nor riches, nor the death of your enemies, and you ask

for it intelligence to exercise righteousness, 12 behold, I will do according to your word. I will give you a wise and intelligent heart, so that there will be no

one before you and that we will never see no one like you"

With this promise from God, we see that the writings of Solomon must be searched again in our time.

The correspondences made here are incomprehensible to non- astrophysicists. This is to say that the basic understanding of the first chapter of Genesis should be reserved for scholars and as we have said, this is why since the beginning of biblical history, it has never been the subject of a sermon or an interpretative publication. The Moon Couple had to appear for the veil to be lifted by the collection of "The verses of the learned" books.

Indeed, by replacing the abyss with the black hole or by the deformation of the tissue of space-time, astrophysicists directly understand what is written in Pr 8 : 26-28 above. Let's confront these sentences with a phenomenon recently observed in the Universe to see if there is a difference, or if the difference is simply due to the language :

"The reason for detecting a jerk in V404 Cygni is the accumulation of material around the black hole ... until it reaches a tipping point, causing sudden bursts of light."(Cfr : - 07/07/2015- Xavier Demeersman, Futura- Sciences : V404 Cygni : when a stellar black hole wakes up)

Ge 1 : "2 ... darkness was over the surface of the deep, and the Spirit of God was moving over the waters.3 God said : Let there be light! And there was light. "

Now let's mix up these 2 sentences that confirm everything we have just said about the behavior of a black hole or biblical abyss :

ITEM	SCIENCE	BIBLE (Ge 1 :2-3)
01	Theaccumulationof material around the black hole.	There was darkness over the abyss
02	As material accumulated in the black hole until it reached a tipping point,	the spirit of God moved above the waters (in the abyss)
03	causing sudden bursts	of light And the light was

For a professor having an objective judgment, he will know that it is the same phenomenon except that the second description (Ge 1 : 2-3) is made by a layman in scientific language and not in cosmology, tan-say that the first is made by a specialist in the field. Certainly for theologians it is almost black and white, but scholars understand and analyze what has just been said. An event narrated by two different people, although the sentences are different, but the understanding of the condition will be the same. Hence these sentences are almost equivalent.

Thus, the understanding of the history of biblical creation is an absolute necessity so that the courses of the theoretical introduction to cosmology or astronomy and The Divine Principle of the Moon couple are part of the courses in the theological and philosophical faculties combined. How can you consider yourself a theologian or a philosopher without understanding the basic ideas about the creation of the universe, the work of our Heavenly Father? Didn't the earliest theories about the universe emanate from the philosophers and theologians of old?

Our wish is that the Principle of Creation of Rev. Dr. Sun Myung Moon becomes a course in the faculties of physics, to understand the source of energy and its role in creation. So the scientists will stop chanting to us that it is "the accident of nature", and will understand the source of consciousness. Note that this branch has nothing to do with an ideology. It is part of the research carried out by the Rev. Dr. Sun Myung Moon by unveiling God's plan for the universe and primarily for man. However, the Rev. Dr. Sun Myung Moon used certain Bible verses, as I do in this book, to demonstrate that this is not a philosophy, but a divine pattern.

Once the source of energy is understood in the different giving and receiving relationships as taught in Creation Principle of Dr. Sun Myung Moon, science will take a big leap forward. Even if the generation of today shuns it, the generation of tomorrow will apply it. By the way, it's already applied.

A last strong argument which passes the biblical abyss for a black hole, as we have restored in the first school of the Universal Cosmological Model, we owe it to the physicist Italian Gabriele Veneziano ; the Creator, in 1968, of string theory, here is an excerpt from his speech :

"In my scenario, we start with an almost empty universe, very, very diluted with very weak interactions between the components. In this near vacuum, however, there are instabilities that can become catastrophic and lead to gravitational collapses like a black hole. And the Universe as we know it would emerge from this initial near-vacuum through this black hole. To be able to explain this phenomenon, we must use string theory and solve its very complex equations which involve a 10-dimensional world. For this we need new results in mathematics. To give birth to a Universe like ours, we must start from a black hole that would have at least the size of an atom nucleus (ie 10 exponents-15m) ". (Gabriele Veneziano, interview by Guy DUPLAT)

http://www.lalibre.be/archive/qu-y-avait-il-avant-le-big-bang; published on 12/15/2005.

This Interview explains the beginning of the universe according to the Bible in a very scientific way, until ending with the notion of primitive atom of the Belgian scientist, the abbot George Lemaitre. Extrapolating with the Bible, we will have the following similarities :

ITEMS	**Gabriele Veneziano**	**Moïse** (BIBLE :Ge 1 :2)
01	Almost empty universe	The earth was formless and empty
02	gravitational collapses like a black hole	Abyss
03	a black hole that would be at least the size of an atom nucleus	A wind of God was moving above the waters
04	the Universe we know would emerge from this initial quasi-vacuum through this black hole	Separation of water from water (Big-bang)

In the light of the above, we conclude that the biblical abyss is not another phenomenon than the deformation of the fabric of space-time, caused by the original material before its condensation under the action of the spirit of God (The First Universal-saddle Force), which moved above the waters (gravitational force). In other words, these verses explain physical action and must be scientifically recognized. The Bible speaks of an abyss followed by the movement of the spirit of God on the elements (waters) that it was able to gather and condense by creating, according to Gabriele Veneziano, "*a nucleus of an atom (ie 10 exponent - 15 m) to give birth to a Universe like ours*".

(Cfr :Gabriele Veneziano, interview by Guy DUPLAT on www.lalibre.be/qu-y-avait-il-avant-le-big-bang; published on 12/15/2005).

Compared to the black holes that form after the collapse of stars, here it is the axis where the original materials are concentrated; what science calls the condensation of the Universe, becoming dense, digging the first deformation of the fabric of space-time, as the great scholar Albert EINSTEIN called it. In relation to the Bible verses, "The earth was formless and empty; there was darkness on the surface of the abyss, and the Spirit of God moved above the waters", we understand that the universe did indeed have an axis and that is where the abyss.

3.3.3. Scientific evidence of the abyss Bible as a black hole

Ge 1 : 2-3 "The earth was formless and empty : darkness was over the surface of the deep God said : let there be light! And there was light ".

As we said at the outset, the abyss is defined, according to the Free Bible glossary, as "*a mysterious world of the primitive ocean, place of darkness and evil powers, of distress and perdition.* "

If such a place was found in space at the very beginning of the universe, what name would we give it today? A black hole. And if 1,500 years before Jesus Christ, someone wanted to talk about something in space, knowing that if we go beyond our horizon, what scientists call the singularity of the black hole, we are lost forever and, in addition, we do not know the mysterious phenomena that take place inside, what name will he use? Abyss.

Even to this day, science cannot explain what goes on inside a black hole. Scientist Stephen Hawking explained at least in very complicated scientific terms from which emanated the light that could be seen on its surface. This Light called "Hawking Radiation" did not come from the depth of the black hole. (Cfr : lemonde.fr, what the work of Stephen Hawking has brought to physics, by Gary Dagorn, 03/14/2018).

If the one who wrote the Bible was oblivious to the heavenly phenomena, he should speak of a hole, but knowing that it is a hole with active and mysterious phenomena within, he spoke of the abyss. It was not until 1967 that the American physicist John Wheeler called this kind of black hole phenomena, to describe a concentration of mass / energy so compact that even photons (light) cannot escape its gravitational force. This is what was considered in the year 1,500 BC, like a "mysterious world, place of darkness and evil powers, distress and perdition", which meets the definition of the abyss.

I have discovered several articles testifying that the Big Bang originated from a black hole. Although I do not share what their authors say about what happened before, but, at least, they confirm my scientific rendering of the schema of biblical creation that the abyss is equated with a black hole. The last article I discovered appeared in an American scientific journal.

All of the following comments could be read
in :https://ingeniumcanada.org/innovation/fr/recit/
un-trou-noir-lorigin-of-universe.

Indeed, according to a hypothesis by three researchers from the Perimeter Institute of Canada, published in the journal "Scientic American", our universe could have emerged from a black hole contained in a universe with a greater number of dimensions. This idea, according to the journal, is based on solid and verifiable mathematics. (Cfr : The Black Hole at the beginning of Time [the black hole at the origin of time]).

These three researchers below, without knowing it, have scientifically confirmed by their research, the biblical scheme of creation. They are Niayesh Afshordi, associate professor at Perimeter Institute, Robert Mann, affiliate member of Perimeter Institute and professor at the University of Waterloo, and doctoral student Razieh Pourhasan. If it hadn't been for my scientific rendition of the biblical creation schema, they would never have known the true importance of their contribution to the Universal Cosmological Model.

According to their scenario, our universe was born when a star in a four- dimensional universe collapsed into a black hole (biblical abyss). For them, our universe has never been inside a singularity; rather, he was born outside a horizon of events, protected from the singularity. It was - and remains - a feature of the wreckage resulting from the implosion of a four-dimensional star.

Always the same papers, their model seems to solve long-standing cosmological problems and - crucially - it produces verifiable predictions.

I hope I have already said that understanding the biblical abyss is very crucial to understanding how the universe came into being. These scientists have joined me. What is good, what are the limits of these authors, and that through the following paragraph :

"We have no idea what a four-dimensional universe might look like. And we don't know how such a "parent" universe came to be. But, researchers believe, our fallible human intuitions are part of a three-dimensional world that may only reveal the shadow of reality."(Cfr :https://ingeniumcanada.org/innovation/fr/).

The other pre-Big-bang (PBB) theories are those of great scientists Veneziano – Gasperini-Damour. Their theories suggest that our Universe arose from a previous phase of contraction of a portion of another Universe. These theories would be similar to those of loop quantum gravity in the framework of pre Big-bang approaches with superstrings. *(Cfr : futura sciences.fr/- Black holes-are they from new universes, L. SACCO, published on 11/05/2007).*

According to these theories, in the pre Big-bang our Universe was created in a certain type of black hole (Biblical Abyss). Once the collapse reaches a limit density for material and limit curvature for space-time, it changes into expansion (Bible's Separation of the Waters from the Waters).

The common thread between scientific pre-Big Bang theories and the Universal Cosmological Model is that the universe arose out of an abyss according to the Bible and a black hole according to these scientists. Other than this commonality, our understanding of before the Big Bang is different, but leads to the same result. For the Bible, at the start, there was a vacuum. But for them, they do not know where the vacuum started, because they ignore the Principle of Creation of Rev. Dr. Sun Myung Moon.

For them, it is a star in a three-dimensional universe that has collapsed, becoming a black hole. However, for a star to collapse, it often goes through a phenomenon called super nova where it emits a very strong light. The Bible also described this phenomenon before the Big Bang with the verse "God said there was light and there was light." In addition, as the star turns into a black hole, not even light can escape. So light and material are all trapped within.

In this case, when the material has been expelled, either by a wormhole or by the black hole itself, it solves the problem of the cosmic diffuse background. Indeed, this light did not appear when the atoms began to merge again, because, that would only have resulted in the formation of a star. Rather, it was a light that had been emitted before and found itself imprisoned first by the black hole or abyss according to the Bible and after by the opacity of material.

The Universal Cosmological Model does not share its point of view on a three-dimensional pre-Big-bang universe. If we stepped back in time, we would have such universes that would collapse and come back, as the American scholar Penrose asserts in his Model of Cyclic Universes.

According to the Bible, the earth (the universe) was formless and empty (Ge 1 : 2) in the beginning before it came to the Big Bang or the separation of the waters from the waters. In this vacuum, according to physics, there was an animation of the particles. These are the so-called particles of water in the Bible; particles which are assembled by the force of a wind from God according to the Jerusalem version, until they form the first abyss of the universe or the first deformation of the fabric of space-time.

After being heated as a result of the condensation, these particles called "waters" ended up emitting their first light corresponding to the biblical account : "God said : there was light and there was light". Was it like a supernova? Shall science give the answer, but not religion. The religion simply says that a light appeared before the Big Bang which has been called "day". This closed the first age of the universe (Ge 1 : 5), biblically called the first day.

And, after an unquantifiable period until then, because it is the astrophysicists who will bring the Cosmological Universal Model into its mathematical phase, there was effectively Big-bang or "Separation of the waters from the waters", followed by after several years of the reappearance of light. This light, in fact, has already been emitted and imprisoned in the past, and will become the cosmological diffuse background.

According to the conclusion of the scientists cited above, this light could come from the star which collapsed and which gave birth to our universe while for the Universal Cosmological Model, this was simply due to the action of the spirit of God on the waters as we can read : "... the spirit of God moved above the waters" (Ge 1 : 2), that is to say the concentration of the original material by a kind of gravitational force.

The contribution of this diagram consists in the statement of the theory, if the scientists were able to collect the material at the zero moment of the supernova, it is the latter which would inform them on what was the material in the Planck era.

3.3.4. The photograph of a black hole confirms the biblical abyss

Black holes are currently located in space. The darkness that reigns there (there was darkness on the surface of the abyss) poses a problem for their location by telescopes. For my readers, I have only been able to explain it through certain phenomena that are unfolding before our eyes, that is, on earth. However, recently, members of the Event Horizon Telescope (EHT) collaboration have indeed revealed the first image of what a black hole looks like and which actually matches what the Bible says : "... darkness on the surface of the abyss... God said : let there be light! And there was light" (Ge 1 : 2-3).

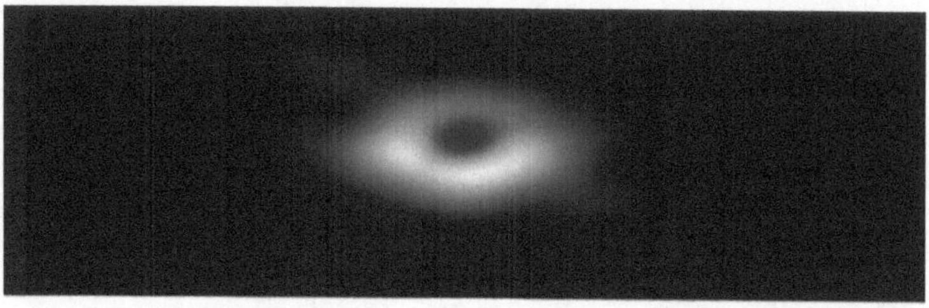

"Astrophysicists obtained the first image of a black hole using observations from the Event Horizon telescope of the center of the galaxy M87. The image shows a ring of light formed by light curving intensely around a black hole 6.5 billion times the size of the Sun".(Cfr : futura-sciences.com/sciences/univers-premiere-image-trou-noir-supermassif, Laurent SACCO, 04/13/2019).

Surrounded by light, the black hole presents itself in its midst as a dark abyss or abyss (there was darkness at the surface of the abyss). Then came the light created by the accretion and ejection of material (And the light was) and outside the disc of light, this is the image of the early universe before the abyss was created (the earth was formless and empty) (Ge 1 : 2-3).

Let's read a comment from the Quora site about the behavior of the current black hole. This comment also answers the question of my readers :

"Why can we be sure that there is a black hole in the photo ? » *We can clearly see an accretion disc (a rotating disc of material) in the process of" falling "towards a massive body emitting violent radiations, which manage to escape because this materialist not yet over- beyond the black hole event horizon. It is still "too far" from the center of the black hole. Normally, such a disc would only form around a super massive body, massive enough to be in fusion and therefore to emit light ... But in the center of the photo, we just don't have enough light so that it is a star. It must therefore be a super massive body, fed with fresh material, emitting no light ... What a black hole!* "(Cfr : Quora.com, am I the only one to think that a photo of a black hole is a bit... im-possible? Response from Chiadai Fouché, Associate Professor of Mathematics at National Education, 05/05/2019).

Being the first to argue that the Biblical Abyss was a black hole, imagine my joy at also being the first in the universe to discover that the photos taken in 2019 by the Event Horizon telescope on the black hole confirm the writings that date back from 1500 years BC. on the origins of the universe. Note that the accretion of the original material in the abyss (abyss) of the black hole is described by the biblical verse "there was darkness on the face of the abyss, and the Spirit of God was moving over the waters." (Ge 1 : 2), that is to say inside.

So, did the universe originate from a black hole (abyss)? If for those who believe in the Bible, there is no longer any doubt, for science, it is a new debate which has just opened and risks disturbing everything and giving reason to the Universal Cosmological Model in confirming the creation of elements. In any case, the Bible has not neglected anything about the components of a black hole (abyss) at the very beginning of the universe as we can read in this press release from the University of Geneva :

"When a black hole is created, it briefly emits a gigantic burst of very energetic light in the form of gamma rays, called a gamma-ray burst. This phenomenon includes all the greatest mysteries of current physics..." (Cfr : Aurélie KUNTSCHEN, 01/15/2019).

This scientific description best explains the verse below :

Gen 1 : 2-3 "The earth was formless and empty, darkness was over the surface of the deep, and the Spirit of God was moving over the waters. God says : let there be light! And there was light ".

In the photo above taken by the Event Horizon team from the center of the galaxy M87, we have as a component which is found in Ge 1 : 2-3 :

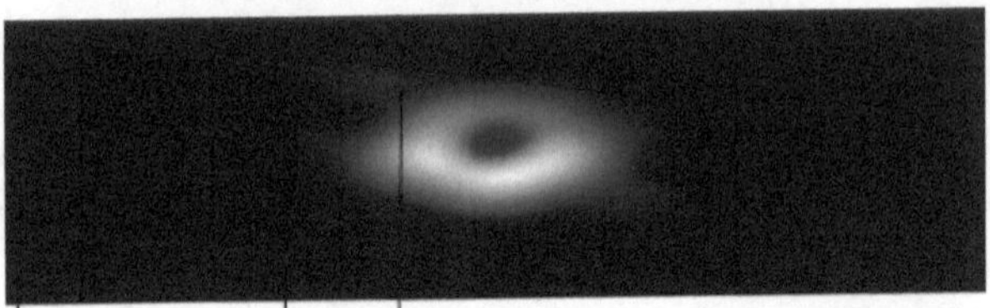

"When matrialfalls towards the black hole, it forms a swirling disc that emits visible and ultraviolet light. This light, in turn, warms the electrons nearby, generating X-rays"
.(Cfr: trustmyscience.com/quasars-revelent-nouvelles-informations-histoire-expansion-univers by Thomas Boisson, 01/31/2019)

– shapeless and empty earth (primitive universe)

– Light (Hawking ray) or information output or stars

– "The darkness on the surface of the abyss where a wind of God moved over the waters" (original material), scientifically known as the horizon of the black hole

– Below, phenomenon not yet photographed and which comes after : And there was light or gamma-ray burst ; also resulting in the ejection of materials

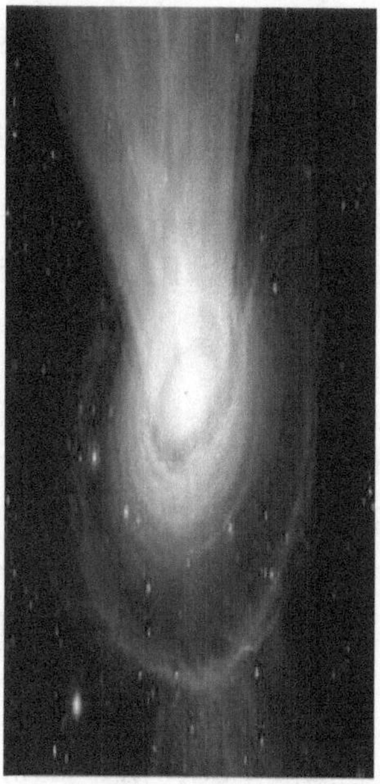

And there was light.
https://cdnfr2.img.sputniknews.com/images/165/96/17.jpg

Pr 8:28 "And let the sources of the deep gush forth with might" or Big- bang.

Everything that has just been demonstrated here as scientific proof of an abyss at the beginning of the universe, it agrees with the description given by Albert Einstein in speaking of the space-time distortion generated by the gravitational fields.

According to Albert Einstein, material creates a distortion in space-time, thus generating, as the Bible says, an abyss. By this we understand that the notion of the distortion of the fabric of space forming the abyss is biblical. If this is the case, it will also mean that to this day we are still operating in this abyss that we call the Universe. Hence the question that French astrophysicist David ELBAZ asked himself : "Do we inhabit a black hole ? And why should we care" ? (Conference to follow on the Rencontres du Ciel et de l'Espace website, Paris 2016).

Who can still say today that the biblical pattern of creation was a tale and a myth ? Who can think of such a scientific scheme in 1500 BC ? It should also be noted that when it comes to predictions, a literal description always offers some differences in concordance with reality. If we accept it for scientists, why should we not for those who did not have knowledge of the current scientific language ?

I might even conclude that Moses did not describe a scientific scheme, but a reality. It is the introduction of variables and parameters, and their restitution into the Universal Cosmological Model that makes it scientific.

3.4. The spirit of God or a wind of God

Ge 1 : 2 "... the spirit of God was moving over the waters" (L.Segpnd Version)

Ge 1 : 2 "... a wind of God was moving over the waters" (Jerusalem Version)

Talking about energy and where it comes from is one of the most difficult points to demonstrate even in science. This is what makes Albert Einstein, a scholar of reference by his famous formula E = mc2 (read "E equals mc squared"). This formula covers part of religion. This point dealing with the spirit of God is going to have to unify religion and science from now on, but, it is a journey.

The in-depth study of the Principle of Creation, branch of the Divine Principle of Rev. Dr. Sun Myung Moon is required for this. Here we will only talk about what is apparent and cannot quench the thirst of scholars because we do not want to go further into the extension.

Moses by the biblical writings teaches us that in order to arrive at the Big Bang, there were physical phenomena of which only the energies could merge (this is studied in quantum gravity or quantum mechanics, a science that studies how material appeared from a vacuum).

Thus, before the appearance of the material world or before the Big-bang, there was the presence of several energies emitting powerful radiations which resulted, under the influence of the spirit of God, the collision of the original particles. (or waters according to the Bible), until you end up with something giving off a strong heat, sometimes defined as plasma.

This paragraph can be understood by this scientific quote :

"The collision between a high-energy gamma ray and infrared light, for example, transforms energy into a pair of particles : an electron, and its antimaterial counterpart, a positron. The same process occurs when medium energy gamma rays interact with visible light and low energy gamma rays interact with ultraviolet light". (Cfr : trustmyscience.com/telescope-fermi-retrace-histoire-rayonnement-stellaire-dans-univers/ by Thomas Boisson, 11/30/2018).

To be properly understood by my readers, plasma, after a symbiosis of the various articles, is a very hot gas (name inappropriate), that is, it forms at extremely high temperatures. At this point, the particles collide so violently that they shatter to form a huge luminous mist that should initially fill the bubble of the before Big-bang. And seen from afar as we said, it can only look like volcanic magma. Hence the use of the word "waters" by Moses which meant at that time, something that looks like a liquid.

First of all, in the following sentence, we directly understand that the Bible speaks of the early Universe :

Gen 1 : 2 "The earth was formless and empty : darkness was over the surface of the deep, and the Spirit of God was moving over the waters",

God himself is Spirit. By speaking of "the spirit of God", Moses meant something that emanates from God. Before the teachings of Rev. Dr. Sun Myung Moon contained in the Divine Principle, the religious knew of God only his attributes (omnipotent, omnipresent, omniscient, etc.). God being Spirit, in Him there is the cause of all spirits. This Spirit of God that Moses speaks of and that moved over the waters is the effect of the First Universal Spirit of God.

This, as I said, is explained in the Moon Couple Creation Principle book. He was the first to discover its mystery and baptized it under the name of Universal Prime Energy (UPE). He explains it as the embodiment of the Word and how it works scientifically in creation.

In the above verse (Ge 1 : 2), we see that the spirit of God was in action above the waters. This is the translation that makes readers of the Bible think there was water below and the spirit or wind of God above it. The spirit in this context is an unseen force acting on the original particles in an empty universe. The scientific terms currently are gravitational force or super force. The waters here represent a form of the original material that scholars seek to discover as we have said.

We directly understand that the Spirit of God represents the Prime Energy which was in action and which brought about the first condensation of the Universe. By default, of name by science, we call this moving force on the original material of the name that the Rev. Dr. Sun Myung Moon had to use, namely, *"the Universal Prime Force"* generated by the Universal Prime Energy which is in God and ask the scientific world to adopt it in order to scientifically immortalize it. Is it a gift ? No.

Indeed, in his concern to explain God and the Universe to intellectuals by words that they can understand and no longer parabolas from 2,000 years ago, the Rev. Dr. Sun Myung Moon had to do special engineering studies in Japan. We have already talked about that.

Thus, the Universal Prime Energy is the fundamental energy of God endowed in the particles. In God, all the attributes, called dual characteristics, are found united. Planned by God and separated, they tend to come back to unity. It can be positive (masculinity) or negative (femininity), which creates the force of attraction between two bodies of the same natures, but opposed (Coulomb's law or of duality which is nothing other than a natural law whose origin is in God). This force manifesting itself in the bosom of God, ensuring his eternity, and in the bosom of the particles for their connection, is baptized by Rev. Dr. Sun Myung Moon who was the discovery, the First Universal force. It is unquantifiable and is different from potential or kinetic energy which requires some external force before it manifests.

The attraction linked to the heart (for lack of the real name) of a male to a female called "love" is linked to this energy and cannot be treated medically, nor calculated in a laboratory.

The Rev. Dr. Sun Myung Moon was the first religious man in contemporary times to explain the eternity, power and omnipresence of God, as well as the eternity of the Universe from a single particle to living beings, in the form of energy. This also brings him scientifically into the ranks of super scientists, because now these writings will pass through the religious realm to join the physical realm and will henceforth be the subject of study.

We have said that, in each being, we find the energy of God called "the first universal energy" which, by the First Universal Force generated by the internal action of giving and receiving, allows the entity to exist and to be active, all forming an energy which has its quantifiable correspondent in the mass.

The spirit of God or a "wind of God" in the Jerusalem version of the Bible, is the energy which allowed material to be put into action : ("the spirit of God" moved above " of the waters") which we call the Universal Prime Force (FPU). Indeed, it is not the Rev. Dr. Sun Myung Moon who directly said that the Spirit of God referred to in Ge 1 : 2 is the universal Prime Force. But it is on the basis of his teaching that We arrived at this conclusion and that I bear the responsibility for it.

Thus, taking into account the curiosity of the scientists who want to know a little more about this name, we expose here some extracts taken from the Divine Principle Book of Rev. Dr. Sun Myung Moon, so as not to tarnish the contents by our syntheses, given the complexity of the matter :

"God, the Creator of all things, is absolute, eternal, self-existent real, transcending time and space. The fundamental energy of God is also absolute, eternal, and self-existent. It is the origin of the energies and forces which allow created beings to exist. We call this fundamental energy of the Universal Prime Energy ... through the action of the Universal Prime Energy, the subject and object elements of each entity form a common basis and interact. This relationship in turn generates all the forces that the entity needs for existence, multiplication and action. Thus, the interaction which generates these forces is called the act of giving and receiving. " (Cfr : Presentation of the Divine

Principle : Pp 29-47, ed. Culture et Paix, Paris).

For science, by "subject" is meant "positive" and by "object", "negative". This is how we find in common, according to the above discussion to which I refer you, that each thing created by God, from the atom to compound bodies, has its positive or negative correspondent, in order to produce a multiplicative result, rather than inertia or annihilation, as we can still read in the same book :

"This is how we discover in all of creation (Universe for scientists) that there is a reciprocal relationship between positivity and negativity. For example, atoms are formed from the reciprocal relationship between positive and negative elements. Atoms themselves have positive and negative characteristics. Based on such characteristics, two or more atoms maintain a reciprocal relationship, forming molecules. Plants contain both male and female elements. Likewise, most animals reproduce by relationship between male and female. All plants and animals exist and are renewed by virtue of myriads of internal and external interdependent relationships of positive and negative elements ... The fact that creation is composed of dual characteristics of positivity and negativity means that God Himself, who is the root cause of all things, is the origin of the dual characteristics of positivity and negativity. (Cfr : the Divine Principle, level 4, p.11).

Through these writings of the Principle of creation, branch of the divine Principle, we understand that in the sentence "the spirit of God moved in the waters", it is about a force on the material of before Big-bang until producing after a long period of the light, which we call according to its corresponding to the current science of radiation (And, the light was). After this radiance which, we have said, marked the final stage of the before Big-bang. The Big Bang must have happened afterwards as we can read in this scientific claim :

"The maximum value of the force of gravity was reached at the time of the Big Bang, that is to say at the creation of the universe, where all material was concentrated at one point and where time did not exist not. It was at the time of this release of energy and light that time appeared ". (Cfr : Claude Sarrailh, Le secret de la vie, ed. Librinova, 2018, P.15).

Without commenting on the author's idea, of which I am unaware of the opinion concerning the front Big-bang, by affirming that the "maximum value of the force of gravitation had been reached at the time of the Big-bang", we see directly that there was indeed an event before Big-bang, as confirmed by the biblical verse "a wind of God or a spirit of God moved on the waters" (Ge 1.2).

This spirit of God, stream from the energy called Universal Prime Force by Rev. Dr. Sun Myung Moon, contributes to the dynamics of the universe as the fundamental force of creatures as an internal force. We find it biblically when God created man by saying in the verse : "God breathed into his nostrils, and man became a living being" (Ge 2 : 7).

Between two separate entities, it takes different names such as gravitation, magnetism, and all the laws of attraction and repulsion known to science. At the scale of the universe, we also find it again in the theory of the unification of forces.

3.4.1. The spirit of God or the primary energy

In the Jerusalem version of the Bible, instead of speaking of "the spirit of God on the waters (the original material), we speak rather of" the wind of God on the waters "(Ge 1 : 2). This means that this energy was found also diluted in material as energy whose physical correspondent is confirmed in part by the famous formula of Albert Einstein E = mc2

[The equation E = mc2 (read "mc equal") was expressed in 1905 by Einstein in the context of special relativity. It means that a mass m particle, isolated and at rest in a frame of reference has, because of this mass, an energy E called mass energy, of value given by the product of m by the square of the speed of light].

So scientists came to the conclusion that energy cannot be created or destroyed. It only passes from one form to another and that everything in the universe is made of energy. This ties in with the "Principle of Creation" of Rev. Dr. Sun Myung Moon.

Scientifically, the Rev. Dr. Sun Myung Moon explains it in the sentences of the Principle of Creation, branch of the Divine Principle, as being the Energy that God gives to every created being. This is in order to initiate within it giving and receiving actions between the positive elements and the negative elements for its survival (individual goal) and for multiplication (overall goal). It is through this Universal Prime Energy in material that we say that everything was created by God and that everything is controlled by these laws, as we can read in the following quotes :

"The Universal Prime Force; it is the fundamental strength of God, the Creator. It is also the fundamental force of creation, the force with which God endows each being or each individual body with truth when it is created as a substantial object of God ... The Universal First Force has its origin in God and constitutes the force that God projects into each being at the moment of its creation". (Divine principle, Level 4, p18-20).

Each particle has within itself the Universal Prime Energy which is the manifestation of the presence of God.

Once entered into action within itself or with another particle, their energies generate the Universal First Force. As within each atom we find protons and electrons, we can say that the universal Prime Force is found and unfolds within each being allowing it to exist as we can read it again :

"*Thanks to the Universal First Force, the action of giving and receiving is directed by a goal of unification and, thanks to its reciprocal organic relations, generates the forces necessary for existence, reproduction and action of things, from the smallest to the largest* " (Divine principle, Level 4, p 21).

This law of the Moon couple, to which I align myself, destroys scientific thought, according to all the energies around us originated during the Big bang according to the Standard Model. However, by scientifically developing the biblical sentences, we discover that without energy at the start, the Big Bang itself could not take place. This is one of the reasons why all models should give way to the Universal Cosmological Model.

Thus, all natural laws in physics and chemistry bearing the names of men are only observations or discoveries and not creations. For example, it was not chemistry (man) that created water by uniting hydrogen with oxygen. But in order for actions to be initiated from within their core, it took Universal Prime Energy. We can schematize it as follows :

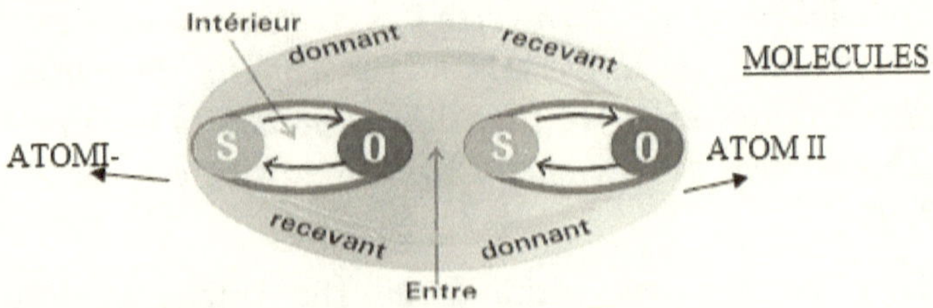

Source : Divine principle, level 4, p.20

According to the principle of creation, as schematized above, atoms are formed from the reciprocal relationship between positive elements (S or subject) and negative elements (O or object). Within each Subject and Object element, we find the same duality, each possessing the primary universal energy, which is in action within it and which allows the entity to maintain and unite with its partner. This holding force within the entity is called the Universal Prime Force.

Then, on the basis of the Universal First Force within them, these two entities S and O will maintain a reciprocal relation, forming an atom. The force which leads to this formation of atoms is called the force of giving and receiving. And finally, it is this last force that causes atoms to unite with each other to form molecules and the latter, on the same force, to form compound or complete elements.

Thus, the Universal First Force, generating relations of give and take between particles, pushes us to attribute all creation to God. Thus, we understand why everything that came from before Big-bang remains puzzling for science and accompanies the universe to its destiny such as darkness (black energy), the spirit of God (Universal Prime Energy) and light (cosmic radiation). All contained in a biblical day of creation or the first day.

We note that compared to the scientific principles in force, the Principle of creation enters much more in depth to explain the forces which animate the particles whose effects quantum mechanics are currently observing.

3.4.2. Positivity (+) and negativity (-) within entities

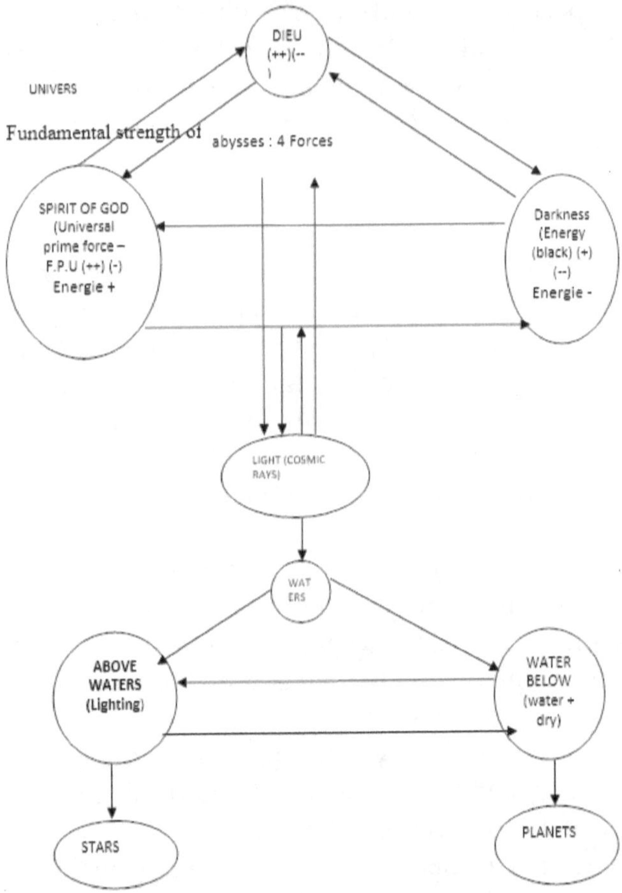

WATER (Particles) and (visible particles)

Ge 1 : 3 God said : Let there be Light And there was light

Ge 1 : 6 Big bang : Separation of the waters from the waters

Ge 1:14

In the diagram above, according to the Principle of creation of Rev. Dr. Sun Myung Moon to which I align myself, we see that God has, in him the two inner traits of positivity and negativity in such harmony that they cannot be separated. His whole creation inherited his characters. Positivity is found in the pro-tone; and negativity, in the electron. However, in these two particles called nucleon, there are, despite their dominant charge, the two characteristics, positivity and negativity, which are at the origin of their continuous rotation, contributing to the dynamics of the universe. This rotation, started at the particle level, is called "spin".

We know that the atom is made up of smaller particles : it is made up of electrons (negatively charged particles); and a nucleus made up of protons (positively charged particles) and neutrons (without electric charge).

These united particles form the atom, the smallest part of a body, and transmit this divine dynamic to it. The atom in turn transmits it to the material which constitutes the universe. This causes all bodies, including the stars, to be composed by the characteristics, which make them rotate around their axis, forming the circular motion. Since the orbit of this movement is constantly changing and influencing the partner star, it creates an elliptical movement. Thus, all celestial bodies, including the earth, perform in circular and elliptical motion and their finite mode of existence in three-dimensional spherical motion.

We can understand celestial phenomena from terrestrial phenomena. Anything spinning in the void without a motive force can never hold a fixed position. It always follows an orbit. This is the case of tornadoes and hurricanes in the land; and, heavenly bodies in heaven.

The dynamics of movement of the Universe are observed by scientists through the writings below, of which they fail to give a convincing explanation :

"*Without claiming that our Universe is running smoothly,*

we have to admit that everywhere, from the closest to the most distant objects, from the most modest to the most gigantic, we observe a great frenzied waltz : specific rotational movements of planets, stars, galaxies, movements of revolution of these same planets, double stars .»(Cfr : H. GIÉ : Angular momentum across the Universe * BULLETIN DE L'UNION DES PHYSICIENS P.427 N ° 712,
Paris).

Here is the diagram :

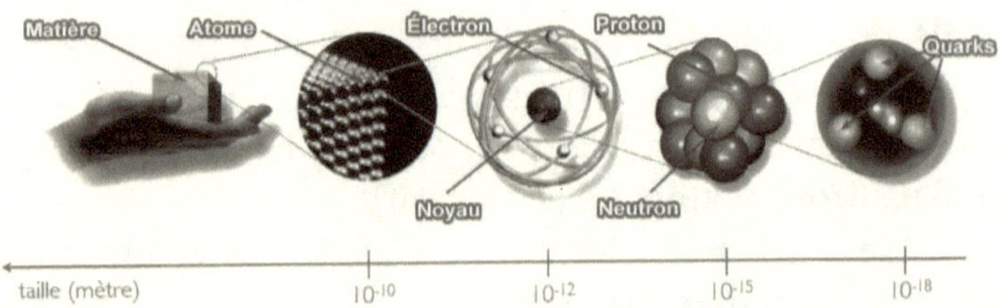

Material is made up of atoms, which are made up of electrons, protons, and neutrons, which are made up of quarks.

Source:CERN/vulgarisation-scientifique.com/wiki/Pages/Les_particules_par : Guillaume Lumin and Franck Stevens. Updated 09/17/2015.

The cause of the dynamism of the universe goes back to God himself, transmitted to particles and to the atom. The Eternity of God himself comes from these dynamic, albeit invisible and spiritual, structures. Projected into the original particles, their tendency is to return to unity at all times. This is governed by the law of ODU (origin-division-union) contained in the Principle of Creation of Rev. Dr. Sun Myung Moon. Synthetically this law says this :

"The dual qualities (subject and object) of God who is invisible, being a subject, generate substantial beings, subjects and objects. Through the action of the Universal

Primary Forces, the subject-object pair forms a reciprocal basis and establishes an action of giving and receiving. The step that precedes the act of giving and receiving is to establish a reciprocal basis, creating an evolving unity. This union is the fruit or effect caused by the action on the subject and the object of the goal element within the Universal Prime Force. In short, the action of ODU is the course of the development of the force which begins in God (the origin) divides and unites again ". (Cfr : The Divine Principle, level 4, p.24).

Schematically, this is symbolized by this drawing

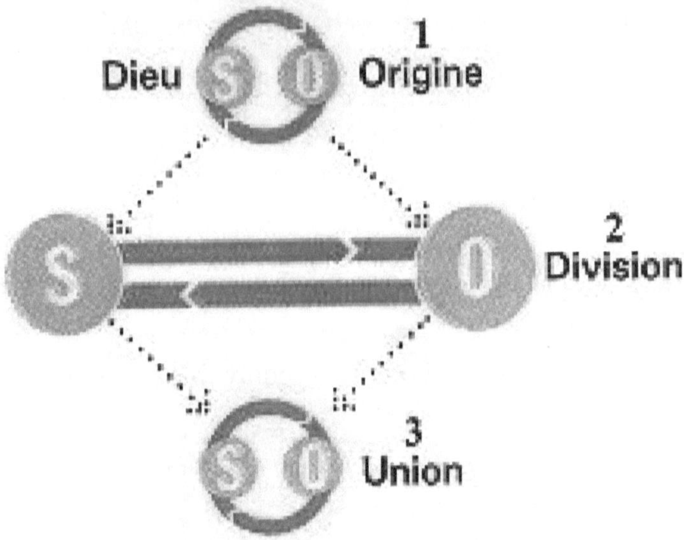

Source : The Divine Principle, level 4, p.24.

As we have said, this divine dynamism transmitted to the level of the particles creates the dynamics of the particles, going up to the creation of the elements (evolutionary union).

This is the greatest conundrum that has just been swept aside regarding the dynamics of the Universe by Rev. Dr. Sun Myung Moon. For more details, we refer you to the book of Presentation of the Divine Principle (ed. Culture et Paix, Paris, Pp 20-42). Summarizing it or extracting a quote would be incomplete.

A static Universe at the start, as Albert EINSTEIN wanted to claim, cannot generate or engender anything. It is its dynamism, also spinning around an axis, there was the abyss and the creation of the primitive particles which will result in the Big Bang. With that, I referred you to the science documentary "New Theory of the Unified Field : Explanation of the Dynamic Universe" on You Tube, when I returned the meaning of darkness.

3.4.3. Meaning of (– –) (+) and (+ +) (-)

Through the signs (- -) (+) we want to talk about a dominant negative compared to a latent positive and a dominant positive (+ +) (-), compared to a latent negative contained in the particle interior. Thus, all entities have a duality of which one force is active and the other latent.

In light of the above, we can say that everything created contains positive and negative polarity, but one of which is latent. This is what justifies the dynamics of the universe.

Scientifically speaking, these points are confirmed by the composition of the proton and the neutron as we can read :

"The first generation of quarks constitutes "ordinary material"; neutrons (of zero electric charge) are made up of two Down (- -) and one Up (+) quarks, and protons are made up of two Up (++) and one Down (-) quarks "; [quark down (-), quark up (+)].

(Cfr : Wikipedia : Quarks).

This is how the positively charged elements which seek to balance around the negatively charged partner elements and vice versa, circling around each other until they create compound bodies if they are able to merge. If this merger leads to evolution, it is a positive dynamic. But, if one begins to destroy the other, it is a negative or unbalanced dynamic and if the two collapse, it is a zero fusion or zero dynamic.

Let's see how these (++ -) and (– +) characteristics of genitors affect offspring :

"In most plants and animals, including humans, the nucleus of the cell has two versions of each chromosome - one inherited from the mother and the other from the father. This is why the animal or plant has two versions of each gene. Sometimes both genes are active, but often only one - the dominant gene

- has an observable action. The gene whose effect is masked is called a recessive gene. " (Cfr : Linda gamin, The origins of life, ed. Artis-Historia, Brussels. 1989, P.30).

By dominant genes, we end up with those with (++ -) and by recessive genes, we mean those with (– +). We can say that by (++ -) it remains (+), and by (– +) it remains (-). But, we deal with the characteristics, even the substances. So what is a scientific sign contribution in science is that in the dominant gene there is a latent recessive gene and in the recessive gene there is a latent dominant gene. All of this derives from the dual characteristics of God, as taught in the Principle of Creation of Rev. Dr. Sun Myung Moon.

When I posed the hypothesis of (++ -) and (- --+), it was for the sole purpose of scientifically restoring this Principle of creation, on the dual characteristics of God and of creation. It was also to clarify what is expected of latent femininity in masculinity and latent masculinity in femininity. I had no idea that I had ended up with the charge composition of quarks that I didn't even know existed. I didn't know I had come up with the composition of the hadrons that make up the nucleus of an atom. At that time, the existence of these particles was completely unknown to me. I told you before that I did business science and economics.

These symbols of (++ -) and (- --+) are personal creations that allowed me to explain everything that has just been said. Applied to living beings, this will further increase the number of possibilities for crossing which sometimes leads to gender malformations.

Thus, in every man, we find a latent feminine aspect (++ -) and in every woman, a latent masculine aspect (−+). The opposite caused by myriads of generations of forbidden love relationships result in individuals with genetic disorders.

Even if we are talking about the original particles, the scientific translation of all these literal explanations of (++ -) and (- - +) we find it in this sentence :

"The proton is a hadron made up of three quarks : two up quarks (with an electric charge + 2⁄3) and a down quark (with an electric charge – 1⁄3). The total of the charges of the quarks therefore gives an electric charge to the proton of + 4⁄3 – 1⁄3 = +1; the proton thus has a positive electric charge". (Cfr : https://trustmyscience.com/hadron/#comment-15314, Thomas Boisson, 08/03/2018).

In conclusion, we say that "the Spirit of God" was the force-generating energy that led to the Big Bang as it can be read again :

Ge 1 : 2 "... and the spirit of God moved over the waters."

Understanding that the purpose of giving and receiving actions is controlled by the Universal Prime Force within each entity, we believe that scientists will no longer use the phrase "this is the nature or the force of nature or the chance of nature ", but rather the first universal Energy or the first universal Force, which is non-quantifiable, because it is the link which connects us to God.

To this day, these detailed explanations of what we have just discussed are found in the book of Divine Principle, taught to members of the Unification

Church of Rev. Dr. Sun Myung Moon. So, it is a necessary for physicists to follow this teaching to better ground their theories, otherwise, they will always end up with an unknown X.

3.5. The water

3.5.1. Moses' use of the word "water"

Gen 1 : 2 "The earth was formless and empty, darkness was over the surface of the deep, and the Spirit of God was moving over the waters".

Scientifically, the material before Big-bang called waters in the Bible, after experiencing extreme condensation, its temperature had increased until the first light came out (God says : there was light ...), then came the Big-bang which means explosion of primitive materials or extension of the Universe. A few seconds later according to the Standard Model, hydrogen was formed and 380,000 years later, the reappearance of light marked the process of the formation of the first star where nucleosynthesis will take place, i.e. the chain of creation heavy elements by the fusion of light elements.

The uniformity of the elements contained in the galaxies shows the biblical truthfulness of having qualified the particles as originated in the formation of all that we see today, including man, the water. In addition, the use of the name "waters" by Moses is said to the fact that the original material, wherever it is to be projected into the Universe, migrates more easily into solid, liquid or gas. This is what justifies that it is found almost everywhere in one of these two forms.

In addition, if today we wanted to give a name using a chemical compound or a physical structure to describe the universe, the answer will simply be "hydro", that is, water.

According to the Bible, the original material called "waters" went through three phases :

– Before Big-bang : the spirit of God moved on the waters;

– At the Big-bang : Separation of water from water;

– After Big-bang : formation of the earth and the lights from the waters below and the waters above, coming an unique water.

Science discredited the Bible's pattern of cosmological creation not only because it did not know the scientific interpretation of the words, but also because Moses spoke of "the waters" to denote the original material of the beginning of creation. or before Big-bang. Today our own galaxy is called "the Milky Way", alluding to pudding when looking at its image as photographed by satellites. However, inside there are gaseous and rocky planets including bright stars.

And what will we see if today we photograph the Universe outside of it? It would look like a black bubble with white spots as shown below. No telescope or satellite has yet done so. Because, how do you photograph something that you yourself are contained inside? But Moses was able to do it through his spiritual laboratory in Midian.

So, speaking of water as a mass before Big Bang, it has nothing to do with liquid water such as the Milky Way that has nothing to do with milk.

We can even say that the water here represents the unknown elements in the mass or the energies that caused the Big-bang.

If we photograph the universe, it would look like a black bubble with white spots like this illustration. And who on the outside could guess their contents without a telescope? Praise be to God for revealing it to Moses.

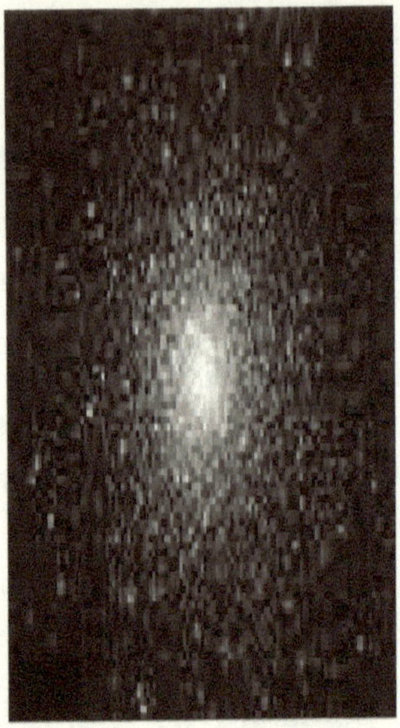

3.5.2. The whole universe is just water

Gen 1 : 2 "The earth was formless and empty : darkness was over the surface of the deep, and the Spirit of God was moving over the waters."

*Ge 1:16. (*From the waters above) *God made the two great lights, the greater light to govern the day, and the lesser light to cover the night; he also made the stars*

Why did Moses had to use the name "waters" as the final stage of these original pre-Big Bang particles ? Admittedly, he is not here to answer this question with precision, but what we put forward as restitution is not far from what he should answer us.

Previously, before the emergence of scientific names, any liquid that could not be described was referred to as "like water"; from sperm to volcanic magmas. Today, science is beginning to advance the idea that the first visible material was in plasma form. A name that myself, as intellectual thinker, did not know. This is why Moses used the name "waters" as the first visible material of the universe to make it clear to all.

In the documentaries "on the origins of the universe", the scholars maintain that at the beginning, everything was only quarks and gluons, moved and collided with each other like a liquid. So with pressure, even the particles behave as a liquid. Hence the name "water" by the ancient writers.

Also, did Moses know that water was made up of hydrogen and oxygen? Did he know that hydrogen was the most prevalent element in the Universe? Did he know that by breaking down the compounds in water, a scientist will immortalize him?

Indeed, by breaking down water, LAVOISIER called one of the elements hydrogen; (Hydro = water) and "gene" (generated)which means, what makes water. In addition, this element occupies 85% of the visible Universe.

In addition, not only is hydrogen one of the essential constituents in water, that is to say one of the molecules vital for life, it is also the most abundant constituent of stars such as molecular clouds. So, Moses was indeed right to use the term water which did not necessarily represent liquid water, but the original materials.

All astrophysicists agree that hydrogen is one of the elements that form outside of stars. In fact, this is the element that is the base of their training. Moses could only use the term "water" just as science today began to use the word ocean without actually referring to our earth's oceans. This is the case with the ocean of mud.

What we will also need to know is that the scientific names of chemical compounds did not exist at that time, 1500 BC. and liquid hydrogen looks like molten steel. The same is true for plasma found throughout the Universe where the temperature reaches extreme degrees, turning certain gases into liquid. This is what justifies the word "waters" by Moses, of which I cannot confirm with precision, nor even science, the state of the original material.

Another proof also of the justification of the word "water" is the presence of hydrogen compounds (which generates water), and that it is also found almost everywhere in the Universe in its three forms ; either liquid, solid, or gas. The shape depends on the temperature to which it is subjected. For this we can read this scientific claim :

"Water is plentiful in the Cosmos. You could almost say this one is "wet".

But, it is rare to find it in a liquid state".

(Cfr : The challenges of science ; the origin of life - edited by RBA Collecionables, S.A.U. Barcelona, 2018, P.22).

We reproduce here three articles from the Internet proving that water is found almost everywhere in the entire Universe, confirming this claim and justifying the use of this name by Moses to designate the original universe.

3.5.3. The largest water cloud in space

Published July 29, 2011 in Futura Science on the Web.

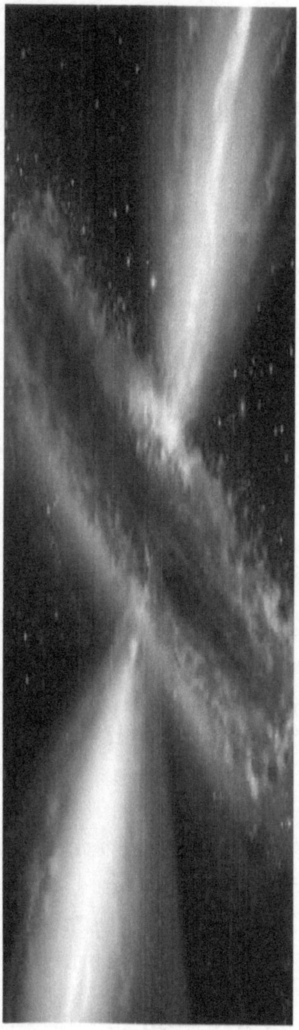

"Scientists have just discovered a cloud of mist, 12 billion light years, weighing 40 billion times the weight of the Earth. It is the largest cloud ever discovered in space. It is in a galaxy 12 billion light years away from Earth that is the farthest and largest water cloud ever observed in the entire Universe,

scientists said. According to observations, made through the telescope on Hawaii's Mount Mauna Kea., The steam that envelops this quasar is enough water to "fill all of Earth's oceans more than 140 trillion times".

Moreover, at the sight of this cloud of mist, in the same article, astronomers even wonder if there was not already the presence of water before the Big Bang, that is to say molecules capable of forming water. Their hypothesis confirms the Bible.

3.5.4. The existence of water in stars

Another example, still taken from futura-science from 07/15/2015, states that even in the stars there is also water :
"The Joseph-Louis Lagrange laboratory (CNRS, Observatory of the Côte d'Azur, University of Nice) observed live and for the first time the transformation of a star of mass near the Sun into a planetary nebula. This star is one of a group of 16 objects known as water fountains. These are evolved stars that show ejections of material that are detected thanks to very intense radiation produced by the water vapor that is in their envelope (the work was published in the Astrophysical Journal of June 10, 2015)" (Cfr : the song of the stars, the new stellar music).

These articles clearly confirm the diagram in Genesis which says that there was a heap of waters (original material) at the beginning, which split into two; the Big Bang, one used for the formation of planets including the earth and the other for the formation of stars. Let us follow these Biblical diagrams :
Gen 1 : 7. So God made the vault and separated the water under the vault from the water above it. And it was so.

Ge 1 : 9. God said, "Let the water under the sky be gathered to one place, and let dry ground appear." And it was so.

Ge 1:16. (From the waters above) God made the two great lights, the greater light to govern the day, and the lesser light to govern the night; he also made the stars.

3.5.5. The existence of water in the planet

A third article which will provide support for thought to scholars on Moses' use of the name "waters" to designate original material in the Bible is as follows :

"It is not only on earth that we find water in the Solar System. Our blue planet, by the way, is far from being the body that has the most, as one would tend to believe. Indeed, water is abundant in comets, asteroids, on Mars and probably also within dwarf planets like Pluto or Sedna. It is also present in very large quantities in a string of moons of the four giant planets. The first ones that come to mind are of course Europe, Ganymede and Callisto, around Jupiter, Enceladus and Titan in the lap of Saturn, Triton in the suburb of Neptune, etc. it is even found at the bottom of the craters of Mercury, the planet closest to the Sun. So where does all this water come from ?

(On 09/29/2014 at 5:34 pm by Xavier Demeersman, Futura-Sciences).

In addition, even in some stars that were believed to be totally devoid of water such as the moon and Mars, we already suspect the presence of ice in parts not lit by the sun. We just recently discovered an article we carry here :

"*And in 2010, NASA's LCROSS mission threw an impactor into a crater near the Moon's South Pole, revealing large amounts of water.*" (Cfr : trustmyscience.com/hayabusa-2-cree-avec-succes-cratere-sur-asteroide-ryugu by Thomas Boisson, 04/08/2019).

3.5.6. The existence of water in black holes

Finally, let's read this quote from physicist Joanne Baker :

"Likewise, radio astronomers measure the speed of particularly bright sources in the center of galaxies, such as H2O masers, which emit powerful radio waves generated by the excitation of water molecules. In several galaxies, the existence of a massive and compact black hole has been inferred from the speed of masers, which follow Kepler's laws."

(Cfr : J. Baker : 50 keys to understanding astronomy, éd. Dunod, Paris 2016, p.141).

Finally, we can say that all the material of the universe can be summed up in one word : "water"; proving Moses right and the Bible.

Although the discovery of the primitive elements designated by the waters according to Moses remains a mystery, nevertheless it is known that those which contributed to the formation of protons and later hydrogen were part of it. At least this is what the scientists claim in their hypothesis which states that the first star was formed without dust as we can read :

"The situation was different when the Universe was very young, and more especially with regard to the very first generation of stars - formed approximately 200 million years after the Big-bang -, firstly because no mechanism cannot be triggered in principle to be invoked for the start of the process initiation, and then because they will have to be constituted in an environment where the only material made available by primordial nucleosynthesis (hydrogen and helium, mainly) is entirely free of dust. "

(The formation of stars : Imago Mundi; Free Encyclopedia, Serge Jodra, 2004 On the Web).

If science can recognize Moses as having lived on earth, he deserves to these days all the awards that are awarded to scientists. And that a Universal declaration be made recognizing God as being the Creator of the Universe because, by reading the story of Genesis, clarified by this book, we discover that these are not predictions but testable scientific truths today.

3.6. The light

Ge 1.3 : "God said : Let there be light ! And there was light".

For a good understanding of this point, I recommend my readers to follow the documentary "Hubble views from space : the work of light" on You tube where we explain at the beginning of the universe, how a light can appear without star.

The idea that a light illuminated the universe before the Big Bang seems absurd to contemporary scholars. For them, the first light only appeared after the atoms had reached a sufficient degree of heat. This is coming after the Big- bang according to standard theory. In the Universal Cosmological Model, on the other hand, the minus one era of the universe, called the first day in the Bible, ended with the appearance of a light. Was it caused by which elements or particles ?

In principle, it is the discovery of this light or radiation before Big-bang that will reveal the elements that were part of the waters of Moses or original material before Big-bang. The future of research into the origins of the universe will resolve this question. My position will take shape on the other hand as I write.

Light is a physical phenomenon, a transport of energy from its emitting source, without transfer of material. So, by saying : "God said, let there be light ! And there was light", we are faced with a physical and not a spiritual phenomenon. For literature, that is to say for the common man, a light is simply a synonym of illumination.

Ignorance of this pre-Big-bang radiation in the Standard Model as in all other cosmological models has led scientists to cast doubts on certain conclusions in cosmology. This pre-Big Bang energy or radiation is not hypothetical like dark energy, and besides, it is active ; hence the name "day" given by God to Moses.

Looking back at fossil radiation, physicists say it is part of the interference we see when we change our TV channels. It is testable and evenly distributed throughout the universe. Indeed, it is biblically recognized that it is the man who gives the name to all that he discovers as we can read :

Gen 2:19 And the LORD God formed out of the ground all the wild animals and all the birds in the sky. He brought them to the man to see what he would name them ; and whatever the man called each living creature, that was its name.

If Moses gave the name of "day" to this light, scientists may have baptized it differently : energies, radiations, waves, etc.

Coming back to our subject again, we said that all light has an emitting source. As for the Bible verse, no one has ever asked the question because it is written : "*God said, let there be light ! And there was light*" (Ge 1 : 3). So, everyone has believed until then, except the scientists, in a miraculous emergence of the light dictated by the word of God.

Today, in the light of this book, we discover that scientists are not wrong in not believing in the emergence of a magical light. This was the result of a physical phenomenon ignored by all, because its source was the contraction of the original material defined as "waters" by "the spirit of God" which played the role of gravity energy. So believers are not so wrong in that everything has been planned by God.

Let's follow that :

Gen 1 : 2 The earth was formless and empty; darkness was over the surface of the deep, and the Spirit of God was moving over the waters. 3 God said, let there be light! And there was light. 4 God saw that the light was good; and God separated the light from the darkness. 5 God called the light Day, and the darkness he called Night. So there was evening, and there was morning-the first day.

This original first light could be the fossil radiation. Why could it be? Because it's up to science to confirm it. We do not want to make statements based on faith in this book. Moses called it a "day." While for the term darkness, we think that it is dark energy and which would be according to the last questionable theory, at the base of the expansion of the Universe from then the Big-bang (Ge 1 : 6-7).

In addition, the gamma radiation that comes from all directions of the sky, despite being, according to science, hundreds of times brighter than a supernova and millions of times brighter than the Sun, only lasts from a few fractions of a second to several minutes. Even if they were also emitted during the deflation before the Big Bang, it cannot biblically deserve the name "day" by their transience.

Also in Ge 1 : 2, we are told as the first step in the first cycle of creation that "*the spirit of God moved over the waters*", which scientifically meant that energy was in action in material, primitive or original. Now, for the Big Bang (the separation of water from the water according to the Bible) to occur, this material had to reach the point where it can have physical reactions producing heat and sometimes fire (the spirit of God moved above the waters) releasing bubbles of energy, illuminating the universe.

The word "light" is biblical evidence confirmed by science that indeed the universe experienced a very high heat phase before the Big Bang. It also comforts us as a demonstration that the creation scheme is not a tale. So, let's see some properties that confirm the appearance of light before the Big Bang.

3.6.1. The properties of the light before the Big-bang

This light appeared before the Big Bang which closed the minus one era of the universe and should carry several properties that may allow scientists to confirm it.

3.6.1.1 Biblical light radiated

Ge 1 : 3 God said : Let there be light! And there was light.
4 God saw that the light was good; and God separated the light from the darkness.

When we see something by the action of light, we say that the light interacts with that object. In this case either the light is blocked or it is deflected, thus creating a shadow area behind the element which deflects it as

in the image below. So our nights are the result of one part of planet earth facing the sun, preventing the other part from receiving light. Thus, the part without the light ends up in the night.

As the earth revolves around itself in 24 hours, after every 12 hours there is a new alternation between day and night. This movement is known as the rotation of the Earth. Sometimes this alternation of time varies with the seasons, caused by the tilt of the earth facing the sun.

The sun on the left and the earth on the right

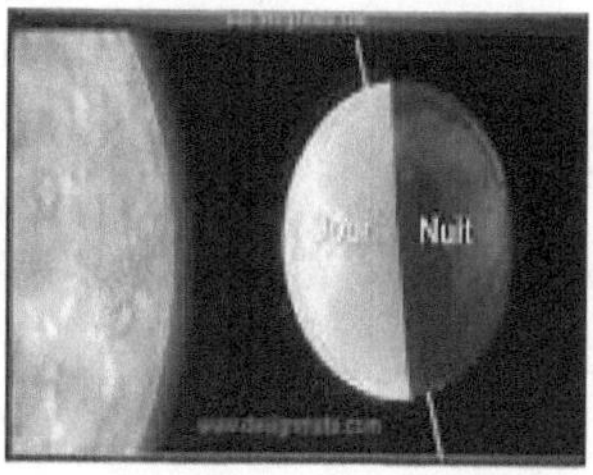

https://i.ytimg.com/vi/4WsXT5inzDg/hqdefault.jpg

Why do believers say it was God who regulated everything? This 12-hour alternation meets the physiological needs of the body related to work, rest and sleep for a good recovery the next day with new energy. This is due :

– into a big comet where no living being can no longer exist.

– The diameter of the earth : a larger diameter of the earth will have many consequences, including increased gravity, which will make the very act of lifting our feet heavier, making us lazy. It will also increase the time of the earth's rotation around the sun with a day going beyond twenty-four hours. A smaller diameter will transform our gait by twists like kangaroos and reduce our days.

At the earth-sun distance : the variation in the distance of the sun from the earth, would transform the earth into a blaze, which will consume all life or There is still a whole infinity of parameters that we cannot quote because of the complexity of these materials such as the magnetosphere, the atmosphere, etc.

This is to say that the diameter of the earth, the position of its orbit and its distance from the sun are calculated by mathematical measures to allow the emergence of life and its maintenance. Who thought of these calculations? For us they believe is God and this belief is confirmed in physics by the Anthropic Principle of Australian astrophysicist Brandon Carter. If he was a priest, no one would ever talk about his principle. One would have defined his conclusions as beliefs.

Coming back to our subject, we can say that our daylight, even nightlight, comes from the sun and has nothing to do with the worm-set below whose light appeared before the creation of the sun, which was created on the 4th biblical day.

Ge 1 : 3 God said : Let there be light ! And there was light... So there was evening, and there was morning : it was the first day.

At the time of this verse, no star had yet been created. Therefore, this light is different from our daylight, as well as in the evening and in the morning. We have already explained this in Chapter 1.

Cosmic radiation (light), produced by the action of the spirit of God (positive energy) on material before Big-bang, should bear the latter's signature.

– In principle, when the light appears, the darkness disappears. In other words, light passes through darkness, and in physics it is said that light has no interaction with darkness. But, here the Bible says "God separated the light from the darkness" that is to say those rays with the energy contained in the darkness. This shows that they always accompany the Universe in its destiny without one annihilating the other. This proves that this was energy radiation and not just lighting.

In scientific language, we can say that this light interacted with the energy that was contained in the darkness and this created their separation. Hence the biblical phrase "God separated the light from the darkness" (Ge 1 : 4). So, we were in the presence of two energies called by Moses "the light" and "the darkness".

– Being created before the Big Bang and projected afterwards, the two energies must cover the whole universe

– The last action before the Big Bang was the separation of the light which had just sprung from the darkness for which the scientists must find a scientific explanation by isolating the sentence "God separated" and by studying the behaviors of the light to through the halo around the disk of each galaxy. If at a certain volume light interacts with dark material, then for us this halo rather represents its imprisonment by this material, which contributes to the dark space. This assertion makes our current knowledge of cosmology still ancient compared to the great intelligence of Moses. We will come back to this in Volume II.

3.6.2. Light, dark material and dark energy

These arguments are in line with the statement contained in the conference documentary "Dark Ideas of Physics" where it is said : "the study of light can reveal the presence of black entities".

In the universe, where scientists have observed the deflection of light by darkness, they directly suspected the presence of dark or invisible material.

Ge 1 : 3 ... God separated the light (radiance) from the darkness.

Thus, in this verse we have the presence of three non-palpable elements : light, unseen dark material and darkness.

The two energies namely darkness and light, both enigmatic energies for science, accompany the fate of the Universe with an element which created their interaction.

If it weren't for this interactive element, even the galaxies would already be undone. And to show that these two energies will henceforth accompany the destiny of the Universe, God by the writings of Moses, called the light, "Day" and the darkness or dark energy, "Night" and that before the beginning of the concept of time. This is to say that these are the names that were given by Moses to these two energies and that it had nothing to do with our dark nights and sunny days.

Gen 1 : 5 "God called the light day, and the darkness he called night".

Why these two great energies for the future of the universe? The answer can be found in the writings of Rev. Dr. Sun Myung Moon who has searched the heart of God and it is up to science to put mathematical formulas into it. This is how instead of para-phrasing it with an incomplete synthesis, I prefer rather to state its answer for this question and which I completely align myself with :

"For a being to exist, energy is needed, and this energy can only be produced through the act of giving and receiving. However, no reciprocity is possible without being at least two. A straight line movement cannot be sustained forever. In order for a being to have an eternal nature, it must describe a circular motion; however, for a circular movement to occur, a give and take action is necessary ". (Cfr : Presentation of the Divine Principle, ed. Culture and Peace, Paris, Pp 42-43)

This non-meditative truth for the religious, is rather of great scientific significance for the astrophysicists, because, it solves the mysteries of the dynamics of the bodies which one observes in the Universe. And, the cause is also unveiled in the same book and we will come back to it. Based on this quote, it is up to scholars to make a parallelism of behavior and existence with all natural phenomena considered eternal.

Indeed, according to science, for there to be the first light, the particles of material collide with each other, heat up and eventually generate gamma rays (light). Thus, these rays emitted, bear the image of the elements that generated them, as we can read :

"Cosmic rays are made up of energetic (primary) particles from space ..."

(School science : cosmic rays www.sciencesalecole.org

We can only explain what happened before the Big Bang on the basis of what happens after the Big Bang, because of being part of the observations.

Indeed, by thermonuclear reactions within stars, the hydrogen atoms by cracking, produce helium while making a jet of light (rays) shoot out. These kinds of reactions confirm what happened according to the Universal Cosmological Model before the Big Bang and that by the verses :

"The earth was formless and empty : darkness was over the surface of the deep, and the Spirit of God was moving over the waters. God says : let there be light ! And there was light". (Ge 1 : 2-3).

For the Universal Cosmological Model, it is the composition of these cosmic rays (God says that there is light), freed of these collected during its journeys, which will reveal to scientists, what history or the material of 'before Big-bang and what the biblical waters contained before the separation of the waters from the waters (Ge 1 : 2).

Given that the universe had to go through a dark period during its extreme condensation up to 380,000 years after the Big Bang, we understand that it will not be easy for scientists to discover and determine with precision the radiance which sprang up before the Big bang, qualified in the Bible by the sentence "God says : let there be light and there was light". It is also not honest to say that this radiance never took place, Moses is not talking about spiritual things in regards to the creation of the Universe (unformed and empty earth).

Physically, he was the one who made all these discoveries (because scholars don't like the word revelation) and wrote it down in great detail without skipping the steps. Only to demonstrate that it was in the action, in the plan and in the will of God, he used the phrase "God said, God made, God created".

To which cosmic rays should we attribute this first light since scientists have discovered several? It is no longer for the Holy Spirit to reveal it to us, for it has already done it to Moses through the Bible. Plus, we said he's the man who gives the right names. Moses nicknamed it "day" for eternity.

Inspired by the source of all the cosmic rays that scientists have already discovered, and not being an astronomer by training, for seven years, I persisted in my writings that it was fossil radiation that filled the whole universe. But, I was disturbed because for scientists, this radiation appeared 380,000 years after the Big Bang. But some scholars say instead that it did not appear, but reappeared 380,000 years after the Big Bang. This is the case with this quote :

"Astronomers know roughly the distribution of material400,000 years after the Big Bang, because that is where the cosmic diffuse bottom was released. "(Cfr : J. Baker : 50 keys to understanding astronomy, éd. Dunod, Paris 2016, p.127).

I rank among the latter. So, for the moment, we are sticking first to the fossil radiation whose observed distribution is roughly isotropic in the sky.

Many may think that this radiation has disappeared

since the Big Bang, giving way to fossil radiation. This cannot be true since God under the writings of the Bible immortalized him with the term "day"; that is, one that has fluctuations that can no longer end. The appropriate scientific term could be "waves".

For darkness or dark energy, we can read this :

"Dark energy remains the dominant component of the universe observable today : it constitutes 69.2% of the mass contained in its volume."

(Cfr : Laurent Sacco, Futura-Sciences, Published on 02/06/2016).

In the sentence "... God separated the light from the darkness" (Ge 1 : 4), hides a great natural law which the scientists will have to study minutely so that their knowledge increases towards that of the Bible. With light representing all that is visible to us, including radiation and quantifiable energies, Moses throughout the Bible says this was separated from darkness, the seat of dark energy.

In the Bible we see that the light was a remnant of the Spirit of God on material(waters). And, by saying that she was separated from the darkness, it demonstrates that the two energies subsist together in the Universe to this day, without one affecting the other, we said. However, since dark energy was once latent, the emission of the first radiation is expected to play a major role in its activity. It might even be the trigger as we'll see in The Ancient Western Models, after we have sounded the death knell for the Big Bang.

In addition, as the appearance of light alternates with darkness, here the Bible says :

Ge 1 : 4 "... God separated the light from the darkness".

This is to say that darkness (dark energy) and light (cosmic radiation) continued their cycle and that these energies preceded the Big Bang and the appearance of the observable Universe.

3.6.3. The origin of the Pre-Big-bang light

This first light, which we say is one of the cosmic rays, was born by the action of what the biblical language defines as "the spirit of God on the waters" by Moses, forming a zone of accretion qualified by the biblical phrase of abyss. And, before the Big Bang intervened, there was first production of this light (cosmic energy or cosmic radiation).

According to the Bible, the light or what we scientifically call the cosmic ray, was issued when the accretion of material formed an abyss by the "spirit of God" (gravitational or deflationary energy) on the original materials of 'before Big-bang (waters). This is an evolutionary process in physics that scientists are familiar with. Indeed, as soon as a black hole (abysses) forms, it attracts on itself the surrounding particles and material defined as water by Moses. These particles collide and generate (gamma) rays. Hence, these scientifically rendered verses :

Ge 1 : 2 "There was darkness (energy) over the surface of the abyss (black hole) and the spirit of God (deflationary energy) moved above the waters (particles and primitive material) ... and there was light (production of cosmic rays).

But to demonstrate that the pattern or the reaction generated was part of the will of God, Moses inserted it with the phrase "God said : let there be light and there was light." By isolating the phrase "God said" as I come to do so in the previous paragraph, we end up with a complete physical diagram on the front Big-bang.

If there is still uncertainty about dark energy and invisible dark material because their explanations are based on the behavior of the visible Universe, with regard to cosmic radiation (light), astrophysicists know that there must be a source or sources. At least, they confirm their physical existences and even manage to define them.

Moses on the other hand, had to confirm the source in the Bible by the sentence "*God says that there be light...*". We have already said that the phrase "God says" was to show that it was in the plan and the divine will. But, this physical light was born on the basis of the elements mentioned before its emergence in Ge 1 : 2 and we have already talked about it.

Indeed, if "*the spirit of God*" was endowed on the matter before the Big- bang (separation of the waters from the waters), God should also foresee the life and the maintenance of all that will happen on the universe after the Big Bang. And among his enigmatic forces for science are darkness (dark energy and dark matter) and light (cosmic radiation), of which it is written in Gen 1 : 2 "God separated the light from the darkness".

According to science, no terrestrial machine or laboratory can produce at the moment, the power of these rays which traverse the whole universe in a record period of time creating uniformity (distributed uniformly in space), so that the sun's rays and the stars lose their intensity as the distance between the street and the street increases.

The previous two paragraphs demonstrate that science still has a long way to go to discover the importance of all that God has created and to recognize its Greatness, to recognize that there was indeed a supreme intelligence that shaped the universe and entrusted it to man (to rule the Universe [Ge 1 : 28]) as his home. But, man is also a component of the Universe, simultaneously breathing air for the maintenance of his existence.

George Gamow was the first to say in the late 1940s that if the universe had really heated up to the Big Bang, the radiation emitted should leave traces. And, it was in 1965 that two American scientists discovered it and science named them fossil radiation or cosmic diffuse background. This radiation will reveal that the temperature of space is strangely uniform even in the most distant places in the Universe that if we take the Big Bang as its starting point, the equation would prove impossible. In other words, according to the Universal Cosmological Model, the maximum warming of this radiation was reached before the Big Bang.

Scientists should know that the stars which are the fruits of the Big Bang behave in something close to what the universe was at its beginning, that is, before the Big Bang. Why is there anything close? Because after the Big Bang, the materials that erupted in the observable Universe, source of the creation of stars or celestial bodies were resulting materials. It is the same when we say that man is created from the dust of the Universe. Thus, man is a collection of resulting dust that we cannot yet decompose to recover the original materials.

Thus in the famous formula of Albert Einstein : $E = mc^2$, if the mass "m" is unknown, all that we can discover there by inverting the formula, it is the quantity of the material contained in this object that is to say the mass of the particles which constitute this object (atoms or molecules) and not the material itself. Only the earth that does it in a natural way by degrading living organisms after death according to the Mosaic law "you are dust and you will return to dust" (Ge 3:19). Law not scientifically popularized because it is contained in the Bible.

In addition, the fact that Moses said that God separated the light from the darkness and that nothing dissipated, it shows us that one of the cosmic radiations would be that light that Moses speaks of. And let's not forget that at that point there was no notion of time. However, Moses closed the first day of creation with the appearance of this light, which scholars are to discover through the analysis of cosmic rays and magnetic fields.

Before the Big Bang, the emanation of light, which symbolized that the first stage of creation had reached its goal after thousands or even billions of years, called a day by the Bible, was the point culminating for the Big bang to take place (Ge 1 : 6 God said : let there be expanse vault between the waters, and let it divide the waters from the waters).

This theory of Moses, using the literary terms of light (radiation), Spirit of God (Universal Prime Force : FPU), of darkness (dark energies), and of the abyss (black hole, dark hole, endless void, space-time distortion), fits exactly with what current science can describe in scientific words and terms. In addition, Moses did not employ a static or stationary character of light and darkness. In verse 2 of Genesis 1 he writes : "There was darkness over the surface of the deep, and the spirit of God moving (was in action) above the waters."

Let's read again how the Bible described the before Big Bang :

Ge 1 : *2-5 the earth was formless and empty : there was darkness over the face of the deep, and the spirit of God moved over the waters.*

God says : let there be light ! And there was light.

God saw that the light was good; and God parted the light with the darkness.

God called the light day, and he called the darkness night.

So there was evening, and there was morning : it was the first day.

Through the biblical verse of Ge 1 : 4 "... God separated the light from the darkness", it turns out to this day that the true dark matter does not absorb the light that is to say the Observable universe. This is how I asked the scholars to study this verse carefully from a scientific point of view.

3.6.4. **Biblical light and fossil radiation**

We said at the beginning that the light that appeared before the Big Bang is the only light that can tell us about dark energy, or even dark material. In addition, the time that science attributes to the appearance of the first lights, 380,000 years after the Big Bang, is too short for a universe which would have to struggle between gravitation and expansion. If so, it simply means that the fossil radiation would be that light that dissipated in the big bang, passed through what is called the dark cosmic age and then reappeared.

We have also said that this radiance arose on the basis of the action of an energy called Spirit of God or Wind of God by the Bible and which resulted in the big bang. These are the equations brought by the Universal Cosmological Model and which have just put cosmology back to the starting point. We cannot do without the Bible for a good start. We will come back to this in the following chapters.

Here are the different extracts from biblical verses (Jerusalem version)

Ge 1 : 2... and a wind of God was moving over the waters.

– God says : let there be light ! And there was light. (Fossil radiation)

5. So there was evening, and there was morning, the first day.

– *And God made the firmament, and he divided the waters which were under the firmament from the waters which were above the firmament.* (Big Bang)

– *This was the second day.*

As we can see, according to the Universal Cosmological Model, fossil radiation preceded the Big Bang.

After having substantiated all our arguments, we offer in this paragraph, an article published in "scienceetavenir.fr" which demonstrates that the biblical diagram of creation is true, and which made that this light reappeared and was not visible according to the science that 380,000 years after the Big Bang. In order not to tarnish the answer, I am asking the author of the article's comments in full, given the relevance of the question as to whether the light had indeed appeared before the Big Bang :

"The Cosmological Standard Model" represents the most likely scenario for the formation of the Universe, which astronomers and cosmologists imagine from data provided by astronomical observations. According to this model, a fraction of a second after the Big Bang (10^{-10} seconds), the Universe was then a very dense inferno with a temperature of almost 10^{15} degrees C. An Universe so dense that particles of light and material were so closely entangled that an outside observer could not see anything with the naked eye.

Since this event, marking the birth of our Universe, 13.7 billion years ago, and as it expanded, the temperature has continued to drop. And in doing so, material organized itself differently with, among other consequences, a release of photons. Based on current models, this release would have occurred fairly quickly. Barely 380,000 years after the Big Bang.

And traces of that original heat and light are still detectable today ...

The presence of this fossil radiation that bathes our entire Universe explains that even in the most remote areas of space, the temperature never reaches absolute zero (that is to say 0 ° K or -273.15 ° C). A theoretical temperature corresponding to a state in which all material is inert and to its minimum energy state".

(Cfr : scienceet futur.fr ; What is the temperature of the Universe ? By Erwan LECOMTE, 09/28/2018).

At the end of the first paragraph of this article, we are told that the temperature of the Universe during the Big Bang was already approaching 10^{15} degrees C. But because of its density, we couldn't even see the light that was emitted by this very dense blaze. Everything was so tightly packed that an outside observer could not see anything with the naked eye. By magnifying, after, 380,000 years after the Big Bang, the photons or particles of the lights were released.

I have already had to criticize the theories which say that it was the fusion of material after the Big Bang that caused light to appear 380,000 years later. I had said that it was very early given the cosmic inflation for material to agglomerate. I think the above article proves me right. Even its author speaks of "barely 380,000 years after the Big Bang". So the light was already there before the 380,000 years, but imprisoned.

In addition, G. Gamow, the American scientist of Russian origin, when predicting the existence of fossil radiation, spoke of the remainder of the light or of the heat which should pre-exist during the Big-bang and not of a light which should start sometime after the Big Bang. So, I stand among the scholars who say that it is only as the Universe expands after 380,000 years that this light begins to become visible.

The last paragraph of the article demonstrates why

this light was defined as "Day" by Moses under divine inspiration through the phrase "the traces of this heat and this original light are still detectable today", and attested by satellites and which causes its temperature to never reach absolute zero.

Let's take a look at this Minkowski diagram of the appearance of light before the Big-bang and its reappearance after the Big-bang :

Conformal Minkowski space-time diagram.

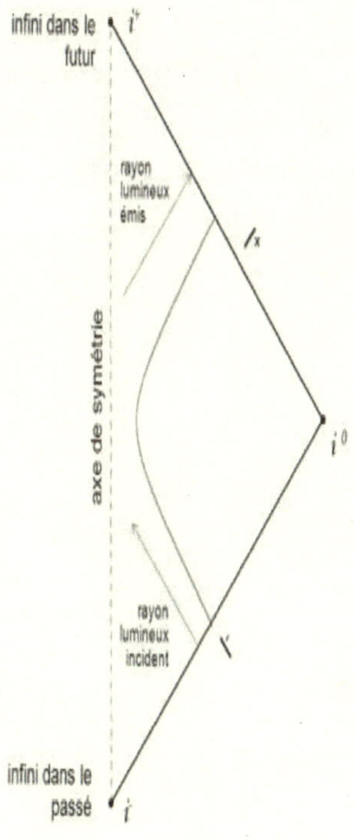

"The conformance of this representation means that a light ray coming from the past follows a straight line inclined at

45 degrees from right to left and that a light ray directed towards the future follows a straight line inclined at 45 degrees from the left to right. The blue curve represents the trajectory of an object coming from infinity in the past and extending to infinity in the future".
(Penrose diagram:Chaours.rv.pagesperso.orange.fr/physique/Stars /Black/penrose.htm).

<u>Our interpretation of the Minkowski Space-Time Diagram</u>

(Penrose diagram: Chaours.rv.pagesperso.orange.fr /physique/ Stars/Black/penrose.htm).
After Big-bang Fossil radiation or reappearance of the first light

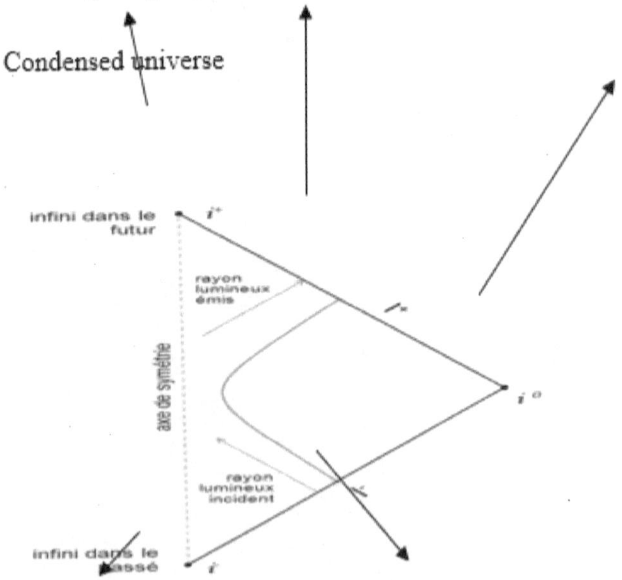

Before Big-bang first light (Ge 1 : 3 "God said there was light and ther was light)

Our interpretation of the Minkowski diagram within the framework of the Universal Cosmological Model, is almost the same as what their authors, Minkowski and Penrose wanted to demonstrate. For us this is proof by scientists that we do not think in a vacuum. This scientific diagram shows that there was indeed an incident light ray before the Big Bang and that in the point **io** where the universe completely condensed, nothing could be visible any more, not even the light

Section 4 : The Scientific Evidence of Before the Big Bang

All the scholars who have spoken of the pre-Big Bang before have been disbelieved because no one could even theoretically prove scientific proof of its existence. Scientists say this was a period when no phenomenon could be explained any longer. If this is true between the period between the time of the Big-bang and that of the Planck wall, but, it did not hold for before the Big-bang.

Indeed, the first five verses of Genesis in the Bible, which is the first book to speak on the cosmos, show us that not only that there was a before Big- bang, but in addition it was a very dynamic period as we can read :

Ge 1 : 2 The earth was formless and empty; darkness was over the surface of the deep, and the Spirit of God was moving over the waters.

1 : 3 God said, Let there be light! And there was light.

1 : 4 God saw that the light was good; and God separated the light from the darkness.

1 : 5 God called the light Day, and the darkness he called Night. So there was evening, and there was morning : it was the first day.

Certainly, those who are not yet convinced will tell me that the Bible verses, although I have scientifically restored all the great sets, cannot be taken as scientific proof of the existence of the pre-Big Bang. Well ! The greatest scientific evidence for the existence of the pre-Big Bang is the Big Bang.

4.1. The Big-bang as scientific proof of the before Big-bang

We haven't yet touched on the Big Bang point. But the pre-Big-bang universe and the post-Big-bang Universe both have the Big-bang in common. Scientists tell us that the temperature of the condensed universe was close to 10^{15}degrees C. No material can scientifically begin existence simultaneously with this temperature. We have said that this temperature was reached following the cosmic deflation of an energetic universe qualified by the biblical sentences of the movement of a wind of God above the waters according to the Bible version Jerusalem (Ge 1.2).

The study of the characteristics or properties of the condensed universe, a theory accepted by the majority of scientists, testifies to the existence of a before Big-bang. And, as the study of our universe begins with the zero moment which is the Big-bang, this word also became the scientific proof of the existence before Big-bang, as we said, through the characteristics of the universe at that precise instant or instant zero. Biblically this zero instant has been characterized by the phrase "God separated the waters from the water" (Ge 1 : 6-7).

Chapter VII : Cosmology According to the Antique Oriental Model

Section 1 : The two basic principles (energies)

.1. **Yin and Yang**

According to the Eastern philosophy, especially Chinese as developed in the book of transformations, we read this :

"God is the great first origin (T'ai ki) of the Universe or origin of the origins. In Him, there is or there are issued the yang and the yin. Thanks to the relationship between yang and yin, the five fundamental elements of the Universe were created, namely metal, wood, water, fire and earth. From these five agents, all other things came into existence". (1).

(1) (Presentation of the Divine Principle, ed. Culture et Paix, Paris, p.28).

Yin represents what is related to femininity or feminine characters. The Yang on the other hand represents what is related to masculinity or masculine characters

My very poor distinction of yin and yang can be extended from the following chart :

Yin		Yang	
black (blue)	黑*hēi*	white (red)	白*bái*
Dark (obscurity)	暗*àn*	bright (luminous)	亮*liàng*
Night	黑夜*hēiyè*	DAY	白天*báitiān*
Moon	月*yuè*	Sun	日*rì*
winter (ice)	冬*dōng*	summer (fire)	夏*xià*
north (cold)	北*běi*	south (hot)	南*nán*
Left	左 'zǔo'	right	右*yòu*
Earth	地*dì*	sky	天*tiān*
Female	女*nǚ*	male	男*nán*
Empty	虛*xū*	full	實*shí*
introversion	內*nèi*	extroversion	外*wài*
Passive	被動*bèidòng*	active	主動*zhǔdòng*
Peer	偶*oǔ*	odd	奇*jī*

Source : "Yin and yang." Wikipedia, the free encyclopedia. Sep 4, 2017.

The explanations of both symbols which represent the two fundamental tenets of Chinese philosophy were developed when the vocabulary was not so diverse, and so, a single word could represent a lot of things in the understanding of its author. Hence, this is what explains that the chart above is not exhaustive.

Thus, for the best development and a good understanding according to our current vocabularies today, the yang represents a positive and active energy, while the Yin represents a negative and passive energy. In this way, we come to the conclusion that Yin and Yang are creative forces.

Today, Yin and Yang are the national symbol of South Korea, and together they call them Taegeuk which signifies the origin of everything in the universe. We understand why this nation whose original name Choson was prophesied by the Bible (Rev 7 : 2) as the land of the rising sun (calm morning) from which the angel will emerge and will mark the seal of God's elected persons. Since the end of the current cosmic cycle predicted by the Mayan calendar, time has come to reveal all of these things clearly.

1.2. Primitive elements according to the Antique Oriental Model

We have said that the biblical "water" in the beginning of the Universe is a mystery because of what it contained. Even science has not yet unveiled the composition of material that erupted in the mass of the Big Bang. For science, even the quarks that have hitherto been recognized as the first elementary particles, were created after the Big Bang during the inflationary period.

So far we have developed the first day of the biblical scheme of creation which corresponds to the creation of the universe. But to this day, esoteric schools and other non-Christian sects, including other Eastern religions, have adopted the Eastern and Western thought of the origin of the Universe as transcribed in the books of ancient Eastern philosophies.

By enriching its language, science for the lack of exegetes tends to abandon ancient wisdoms. Thus, for thousands of years, the ancient writings have been the subject of much philosophical explanation without really penetrating science. It is even taught as such in philosophy and theology.

From now on, the books of the collection "The Verses of the Scholars" comes to dissect the explanations that will enter into the history of cosmology, the ancient Eastern and Western models.

Section 2 : Explanations of the primitive elements of the ancient oriental model

We are not going to analyze deeply the Eastern or Chinese philosophy. What interests us is the hidden explanation of these concepts. For thousands of years, since the existence of these thoughts, no exegete has been able to detect their content root. This is because science was not as developed as it is nowadays. The Bible whose writings dated back to antiquity, by speaking of "the moving of the spirit of God (energy) on the water (material), there was light" the Bible really passes for an exceptional book that no scientist could neither ignore nor deny any more.

Indeed, every physicist knows that the compression or contraction of matter in space causes them to heat up, increases the temperature and emits light and has been responsible for the Big-bang or the formation of stars, or even supernovae.

In light of the Divine Principle of Rev. Dr. S.M. Moon, we have said that in God there exists the universal prime energy which is the source of the energies and forces which allow the created beings to exist in this world. The Energy as a generating force of movements and charges, in God himself this force is called the universal primary force.

In the Eastern religions, the origin represents God, while Yin and Yang are forces or energies that emanate from him. Their relationships generate the primary universal force which is the set of forces that God projects into creation to unite according to their nature, and form composite elements. So, yang and yin, according to the Divine Principle of Rev. Dr. S.M. Moon, are parts of the dual characteristics of God positivity and negativity, masculinity and femininity, spirit and form, cation and anion.

With the development of physics, we can add particles and antiparticles, matter and antimatter.

That is why, according to the Eastern philosophy, it is said that thanks to Yang and Yin came the five fundamental elements (metal, wood, water, fire and earth). According to the same philosophy, yang and yin put together are called the voice (Tao). This agrees with the famous sentence in the Bible that says that "God commanded /said" before each creation action which also represents the way.

This way is explained by the apostle John through the following verses :

John 1 : 1 Before the world was created, the Word already existed; he was with God, and he was the same as God. 2 From She the very beginning the Word was with God. 3 Through him God made all things; not one thing in all creation was made without him. 4 The Word was the source of life, and this life brought light to mankind.

In the light of the Eastern philosophy and the language used by Moses, this sentence according to the first chapter of the Gospel of John is going be announce like this :

John 1 : 1 In the beginning was the Voice, and the Voice was with God, and the Voice was God. 2 he was in the beginning with God. 3 All things were made by him, and nothing that was made was done without him. 4 In him was life, and life was the breath or spirit of God in created beings.

The Book of Transformations and the Bible are limited because of the non- development of science at the time they were written, and so, the language can not only be religious. Thus, the yang and the yin put together form the Voice.

For example, in the sentence : *"Then God commanded :*

Let there be light!
– and light appeared" (Ge 1 : 3).

In this sentence we have the projection of God's energy of yang and yin through different elements (water) which resulted in the formation of light (fires) and, according to Moses, this light was a prelude; as the starting point of the Big-bang which allowed the formation of the visible Universe.

2.1. Water

Ge 1:*2 "The earth was formless and void, and darkness was over the surface of the deep, and the Spirit of God was moving over the waters.*

Just as in Ge 1:2, Moses began to describe the constituent elements of the Universe before the Big Bang as "the waters", although according to science, in the vacuum there were already virtual particles. The author of the book of transformations also mentioned among the primitive elements "water". The Ancient Western Models that we will see in the following points have also designated it by "water".

Thus, as in the Universal Cosmological Model, we draw as the first element before the Big-bang mentioned in the Ancient Oriental Models "water" or "waters" which represented the original particles, the original materials and the original elements.

2.2. Fire

The fourth fundamental element mentioned in the Book, of Transformations as prime matter in The Eastern Model is fire. Today, we know that fire is not a chemical element.

As I said, the author of the Book of Transformations did not take into account the classification for us to judge it. But, in the light of the Universal Cosmological Model, we know that the name used in antiquity represented the states of matter.

Indeed, fire has always been the consequence of something. It is a release of heat and light resulting from combustion. Its flame consists of incandescent particles of the burning substance and certain gaseous luminous products. To burn in the ambient air, a fuel and a sufficient quantity of oxygen (usually provided by the air) or chlorine are usually needed to fuel the combustion. (See : "fire." Microsoft® Encarta® 2009 [DVD], Microsoft Corporation, 2008).

In antiquity, fire was produced either by friction or percussion. To produce fire by friction, two pieces of wood covered with fuel are rubbed against each other until the combustion temperature is reached. Percussion consists of hitting two pieces of stone against each other or hitting stone against certain metals.

However, there are also cases where, if you trap certain compounds, they will also be able to produce fire without being shaken. An example that is within everyone's reach : *let's take pharmaceutical permanganate, reduce it to dust, pour the powder into a funnel-shaped paper with the end attached. Pour a few drops of glycerin into it and tie the ends of this paper to suffocate the air and everything inside. A few seconds later, you have fire. The paper is used as fuel.* This experiment to be done in an open place.

The reaction may be due to the presence of potassium in the permanganate which has the property of reacting violently in the presence of water, producing potassium hydroxide and hydrogen gas, which burns spontaneously.

Fire can also be produced using a lens or a curved

reflector by concentrating the sun's rays on a combustible substance. And, currently, we are witnessing bush fires caused by the high temperature of the sun on certain grasses. Even the lightning that strikes a tree produces fire. Therefore, fire as we know it cannot be cited as a constituent element of the universe.

By scrutinizing some esoteric books, we talk about fire as the main element in the creation of the Universe. But when did fire intervene and in what way ? It was the action of the energy that Moses calls the Spirit of God on the primitive matter called water in the book of Genesis. In the ancient Eastern model, these energies are designated by the action of yang and yin on water. This is how the fire sprang up by emitting the great light mentioned in this point. This brings us to the conclusion that the Bible is the Word of God by excellence.

The fire was produced by the bombardment of the original particles that projected great heat on the surface, while at the same time illuminating the space (God says that light be and light was (Gen 1:3). When the heat reached its peak, there was the bursting of the original bubble (Big-bang).

As we wrote in the introductory part of the Eastern Model, in the Book of Transformations, the five elements were spread out without demonstrating how things could happen and ignorance remained for millennia. But through this book, we now know that from the movement of yang and yin over "water", which was the state of primeval matter, heat (fire) was created. Fire or heat producing light, the Bible must have quoted it before the Big-bang.

If the Bible hadn't quoted the light before the Big-bang, we would be upset. Thank you, Moses, for not neglecting anything. It is as if he knew that one day his scheme, inspired by God, would be scrutinized. This analysis joins the Bible where Moses, after (Genesis 1:2) water, spoke of the appearance of light. Let us read :

Gen 1 : *"2 The earth was formless and void : there was darkness over the surface of the deep, and the Spirit of God moved over the waters. 3 God said, "Let there be light! And there was light. »*

If we superimpose our reasoning from the Book of Transformations and Genesis, it gives us two identical sentences :

Oriental Philosophy	Bible
From the movement of yang and yin over the water sprang fire	"The spirit of God moving over the waters, light came forth" (Gen 1:2-3).

A light never appears alone. In our context here in the Bible and the Book of Transformations, the light was the consequence of the great heat (fire) coming from the movement of God's spirit over the primitive elements called "water" which scholars have always sought to know better. Let us read how Moses describes the manifestation of fire :

Ge 1:3" *God says : Let there be light! And the light was 6. And God said, "Let there be an expanse between the waters, and let it separate the waters from the waters.*

Why did Moses speak of light and not of fire? Because it is fire that produces light, which sometimes manifests itself as radiation. These rays carry the message of the pre-Big Bang information to this day. Hence a re- reading of the image of the cosmological diffuse background which is considered by science as being the first light emitted in the universe imposes itself.

In Ge 1 :2 (there was darkness on the surface of the abyss and the spirit of God was moving over the waters), it is mentioned as primitive components of the Universe, the darkness seat of dark energy, of the abyss that we can qualify as a black hole or the first deformation of the fabric of space-time, of the spirit of God representing the energies qualified as yang and yin, of super- gravity or super-force, as well as the "waters" representing the other original particles.

Fire also represents the expression of Big-bang whose temperature varies beyond $10^{15\circ}$ C.

Chapter VIII : Cosmology according to the Western Ancient Model

As in the Eastern Ancient Model, we do not want to talk about the Western Ancient Model in its historical aspect, but, by deciphering the thought of their authors as we have done for the Bible. To do this we have retained two philosophers who have reasoned a lot about the origins of the universe. It is about Empedocles who left us a sandal as a remnant, meaning that we must follow him. The other philosopher that we classify as the greatest is Aristotle.

Empedocles who was one of the greatest philosophers of antiquity is not as well-known as Aristotle. In the cosmological field, the present book "The Verses of the Learned, Volume I" has just reversed the trend by saying that it was Empedocles who was the greatest, as is Hippocrates (-460 to -377 B.C.) for medicine.

I will make only a brief development to open the key to knowledge, and leave it to philosophers, metaphysicians and astrophysicists to make detailed developments in their books.

In conclusion, I will say that the Ancient Western Model has come again to make a philosophical arbitration between the Standard Model and the Universal Cosmological Model. it is up to the scientists to follow mathematics.

Section 1 : EMPEDOCLE (Approximately 490 to 430 BC)

My choice for EMPEDOCLE comes from the fact that other renowned ancient philosophers have described and analyzed the universe based on what they saw. They limited themselves to the "how" as contemporary science has done, whereas EMPEDOCLE spoke about the origins of the universe, that is, about what we don't see. This requires wisdom and a breakthrough into the infinite. This is not given to everyone. Even the scientists of the contemporary world have put a cross on the origins by proposing different models to us.

1. Presentation of the cosmological ideas of Empedocles

Empedocles was a Greek philosopher, poet, engineer and physician from Sicily, from the 5th century B.C. He was among the first philosophers who tried to discover the origin of the cosmos, its foundation and explain how it works. Even if he is less known than other philosophers such as Socrates, Pythagoras, Aristotle or Ptolemy, in cosmology I consider him to be the greatest. Although my statement may be spontaneous Friedrich Nietzsche (1844-1900), who spent time studying all the great ancient philosophers, had come to the same conclusion.

The originality of Empedocles is that it sets out two principles that cyclically reign over the universe, Friendship and Hatred. Originally, these principles gave rise to the four elements of which all material things are composed (water, earth, fire and air). Friendship, in the cosmological context of Empedocles, is a unifying and cohesive force that makes things tend towards unity (for example, living organisms). Hate, always in this context, is a force of division and destruction that makes things tend towards the multiple. (Cfr : Empedocles : Wikipedia)

Let's first read the EMPEDOCLE ideas presentation :

His physical doctrine makes the four elements (Fire, Air, Earth, Water) the principles of all things. To these Elements are added the Forces of Love and Hate: Love even brings together what is dissimilar, and Hate separates what is joined: ...

At one time, one formed the Multiple, at another time it divided, and from one came forth the Multiple - Fire, Water and Earth and the mighty height of Air.

Originally, all the elements, united by Love, formed a homogeneous mass called the Spherus. (Hatred began to set in) Discord penetrated it and separated the elements from each other. Then, after this period of dissolution, Love penetrates again into the world and takes back to Discord the dissociated elements. It occurs first at a point (either by the action of Love or as a result of a disruption of the balance between fire and air, so that the beginning of the world would be the final result of the disordered movements imprinted on the whole by Discord, a whirling movement by virtue of which some of the substances are mixed: Discord is pushed out. The whirling motion, at first very slow, expands more and more; the separate substances are attracted by the primitive mixture; Discord is increasingly driven outwards. Thus is formed the world, which must perish again when, after complete unification, it will have returned to the primitive state of the spherus".

(Cfr: www.Wikipedia. EMPEDOCLE).

In this very philosophical development, we see that, Empedocles speaks about the origins of the universe by a force called love which united everything by forming the spherus, until its destruction by a Big Crunch or the return to the primitive state of the spherus

This makes his philosophy very interesting for astrophysicists who are reading me now. These writings date from the 5th century B.C.

But, let's first continue with his philosophy about the functioning of the universe :

– The duality and the opposition of the forces of Love-Hate applying to these four Elements is moreover subject to an alternation : to a state where Love alone reigns and where everything is united (the Sphairos), follows the progressive introduction of Hate until the complete separation of the Elements, Love reappearing then brings things back to unity and towards a new cycle «.

– By the action of the sky, the Earth remains quiet by the effect of a whirlwind that surrounds it.

– For they alternately prevail in the revolution of the circle, and pass from one to another, and become great according to the turn assigned to them.

Moreover, it is the philosopher Empedocles who first proposed in the West, a correct explanation of solar eclipses. (Cfr : www.WIKIPEDIA : Eclipse).

– Scientific restitution of Empédocle's thoughts

– The energies of the beginning of the Universe according to Empedocles

Ancient and contemporary philosophers have focused on the four elements cited by Empedocles, forgetting that the most important or the cause are the two great energies he highlighted. Among these energies we have Love and Hate. It is like Yang and Yin in the Ancient Oriental Model. These two energies are described as the basic manifestations of the functioning of the Universe.

This is similar to the name of certain energies which we list in the non- exhaustive tables above and which form all the attributes of the dual characteristics of Go d according to the Principle of Creation of Rev. Dr. SUN MYUNG MOON to which we subscribe :

BEFORE Big-bang : Unification of forces

DIVINE PRINCIPLE	**Object or passive force (negative) and subject or active force (positive)**
BIBLICAL MODELS	Darkness, Wind or Spirit of God
AntiqueORIENTAL MODEL	Yin, Yang
Occidental MODEL	Hatred, Love
STANDARD MODELS (Quantum mechanics)	Energy, super force, supergravity

AFTER Big-bang : Separation of the Forces

DIVINE PRINCIPLE	FORCE OBJECT OR (ACTIVE-NEGATIVE)	FORCE SUBJECT OR (ACTIVE-POSITIVE)
Cosmological model Universal	Dark energy	Wind of God: supergravity
ORIENTAL MODEL	Yin	Yang
ANTIQUE MODEL	Discord - hate	Love (friendship)
STANDARD MODEL	Dark energy	Gravitation and / or dark matter

These non-exhaustive tables further open the understanding of what will follow.

In the sentence "Friendship and Hatred have given rise to the four elements of which all material things are composed : water, earth, fire and air", we see that these two human characters have nothing to do with that of man. In ancient times, there were no scientific terms to describe certain manifestations. This is how Empedocles borrowed these names ; hatred and love from literature to explain the forces (energies) of the beginning of the Universe. These names are currents or energies which we will develop in the following points.

2. **Love (Friendship) - the Spirit of God**

For Empedocles, "Friendship or love is a unifying and cohesive force that makes things tend towards unity (for example, living organisms) or even one when it comes to the cosmos". In science, when we talk about forces, we directly see energy ; here we are facing an active energy, which exerts the force of unity. In science, by its properties, this force is synonymous with gravity, super-gravity or super-force.

In the Bible, this force is described by the phrase "A wind of God, or the spirit of God moving over the waters" (Gen 1:2). This definition of Empedocles allows us to understand how these energies acted before the Big- bang. In today's science, supergravity is more and more spoken of about as the original force that acted in the Big-bang. And, its definition corresponds to that of Empedócles concerning love.

Thus, in the following sentence of Empedocles : "Originally, all the elements, united by Love, formed a homogeneous mass called the spherus", we notice that we are in the before Big-bang. It is this mass which was qualified by the Bible as "waters". Science has never given the name to this mass since they rejected the name of primitive atom of Abbot LEMAITRE and replaced it by "the condensed universe". Moses went into much more detail in describing that this condensation caused a damage or deformation of the fabric of space-time within an empty Universe.

For Empedocles, by a force he called love, since love unites, a homogeneous matter was formed which he called "spherus" and which after the occurrence of another force he called "hate", this mass found itself divided.

Speaking of spherus, Empedocles, without knowing it, joined the bible which qualified this homogeneous mass by "waters". Later, the Belgian scientist Abbe Lemaitre called it the primitive atom.

For the Universal Cosmological Model, it is the confirmation of a bubble before the Big-bang that is important. Therefore, whether it is in the ancient Western model defended by Empedocles, the biblical creation scheme or the Universal Cosmological Model, it is the original particles that were condensed.

Conclusion

Now, the Universal Cosmological Model gives scientists new proofs for thought for the pre-Big-bang cosmos. The Bible has spoken in understanding, and it is opportunity for scholars to extend it more. We noted the following major structures before Big Bang :

– The earth was formless and empty : Universe without form and empty at the start; a physical nothingness

– The Darkness : seat of dark energy, dark energy, void energy. If we remove all the stars in the universe, there will only be a dark space containing the energy of the void.

– Surface of the abyss : contours of the black hole, area of entry for accretion of material.

– Abyss : black hole or first deformation of space-time; place of concentration of the first original particles

– the spirit of God or a wind of God : Universal first force. Energy of material and on material, Pre-Big-bang contractive force.

– Waters : original materials at the start, sets of particles and materials before Big-bang which will evolve into materials after Big-bang.

– Lights : heat, fires, radiation

The combination of these seven sets resulted in the Big Bang or the separation of the waters from the waters (Ge 1 : 6).

Throughout all that has just been said about the before Big Bang, the broad sets about the formation of the Universe have already been discovered by science and described by Moses in the Bible. He intercalated the physical manifestation for the passage from one phenomenon to another with the expressions "God said, God made, God created" to demonstrate that it was part of God's will and plan. Moses limited himself to telling the story, but he did not neglect anything. If he had had the time to offer Israel a nation, all of his formulas would have been exposed and taught.

I spoke a lot on the Big Bang before because I was also the first to speak with conviction and the first to discover in the Bible cosmological phenomena that science is confirming, namely :

– The Universe had a beginning (in the beginning God created the heavens and the earth);

– In the beginning there was chaos (the earth was formless and empty);

– Before the Big Bang was an active time (the spirit of God moved over the waters until the light appeared);

– Dark energy is biblical (there was darkness over the surface of the abyss)

– The deformation of space-time or the black hole is biblical (the abyss)

– The Universal Prime Force (the spirit of God or a wind of God moving above the waters) with gravitational properties;

– The particles or original materials are biblical (the biblical waters before their separation)

– Fossil radiation is biblical (and, there was light)

– The manifestation of dark material is biblical (and, God separated the light with the darkness)

– The Big bang is biblical (God separated the waters with the waters).

The other findings are in the preceding volume. It was not after reading cosmology that I started to research these biblical phenomena. Rather, it was after discovering these phenomena that I started looking for their correspondent in cosmology so that we can speak the same language with science.

BIBLIOGRAPHY

I. BOOKS

– Presentation of the divine principle : ed. Culture and Peace, Paris, 2013

– The Divine Principle, level 4, introduction

– J. Baker : 50 Keys to Understanding Astronomy, ed. Dunod

– BABYLON THE GREAT FALLS - the kingdom of God began its reign, ed. Watchtower, JW.org

– Claude Sarrailh, The secret of life, ed. Librinova, 2018

– AurélienBarrau : Big-bang and beyond, stroll in cosmology, published by Dunod, Paris

– P. Batchelor : Energy, sound and light, ed. New Horizons - ARS, Paris, 2016

– Damian Grint, Mysteries of the Universe, ed. Christophe Colomb, Paris, 1985

– Linda kid, The origins of life, ed. Artis-Historia, Brussels. 1989

– Melvin and G. Berger : What makes the ocean ripple ? ed. New Horizons

– ARS, Paris 2012

– J. Herrmann, Atlas of Astronomy, ed. Stock, Paris, 1976

– The challenges of science : the origin of life, edited by RBA coleccionables, S.A.U. 2018, Barcelona.

II. ENCYCLOPEDIA

– Geographic exploration. "Microsoft® Encarta® 2009 [DVD]. Microsoft Corporation, 2008

– Joe Bergeron, Encarta ® 2009 Modeling a black hole

– Dicosencarta 2009
– Moses. "Microsoft® Encarta® 2009 [DVD]. Microsoft Corporation, 2008
– Wikipedia ; Metaphysical
– Wikipedia ; Elementary particles

III. **REVIEW**

– Science & Vie Special issue n ° 288, 09/2019,

IV. **PRACTICAL WORK**

– MPANZU K.B. and NIMI M., explanation of quantum gravity, National Pedagogical University (RDC), 2017, unpublished

V. **WEBSITES**

– At the origin of the Big Bang : Abbé Georges Lemaître (Humanity ; Science and religion)
– vinsvignesvignerons.com/Geologie/Une-breve-histoire-de-la-planete- terre
– Mark Eichenlaub in Quora : What did Richard Feynman mean when he said, "What I cannot create, I do not understand» ?
– futurasciences.com. First stars in the universe : signals finally detected, Laurent SACCO. The 02/03/2018
– trustmyscience.com/telescope-fermi-retrace-histoire-rayonnement- stellaire-dans-univers/by Thomas Boisson, 11/30/2018
– Dr. David P. Stern Mail to Dr. Stern : education ("at" symbol) phy6.or
– astronomes.com/le-big-bang/particule-virtuelle/ August 9, 2017 by Oli- vierEsslinger
– matierevolution.fr, Dark matter, dark energy : the missing link ?
11/13/2007, by FaberSperber, Robert Paris : fr.sputniknews.com : a black hole emits mysterious sounds

– www.techno-science.net; Successful test of Einstein's theory of relativity near a supermassive black hole by Adrien "07/27/2018

– https://ingeniumcanada.org/innovation/fr/recit/un-trou-noir-l origin-de- lunivers.

– scienceetavenir.fr; What is the temperature of the Universe? by Erwan LECOMTE, the 09/28/2018

– monde-diplomatique.fr/2012/07/VOILLIOT/47999 - summary taken from the book "Man according to Karl Marx by YvonQuinioun

– Anne-Laure Lebrun, Figaro.fr, published on 02/05/2018

– F. J. Dyson, Infinite in all Directions, Harper and Row, 1988

– annuairediggdes.coursgratuits.net/c7_bigbang/p715_partvirtuelles.phpparOl ivier Esslinger

– The force that comes from the void - Astrid Lambrecht in monthly n ° 376 /06/2004, p.4

– H. GIÉ : Angular momentum through the Universe * BULLETIN DE L'UNION DES PHYSICIENS P.427 N ° 712, Paris

– Astrophysical Journal of June 10, 2015)" : the song of the stars, the new stellar music

VI. **SCIENTIFIC DOCUMENTARIES**

– "The Origins of the Universe" by Prof. Dr. A.E. WILDER Smith

– Cosmos - Beyond the Big-bang »written by Matt HICKEY

– Hubble views from space : the work of light

– Quantum physics for dummies

– "HAS THE UNIVERSE EVER BEGINNING? "Conference by ETIENNE KLEIN

– Beyond the Big-bang

– New theory of the unified field : explanation of the dynamic universe

– "New Theory of the Unified Field (Explanation of Gravitation)

– The Mysteries of the Universe

– "The end of the universe

– "The Unexplained Phenomena of the Universe" written by Laura Verklan Armstrong

– The Milky Way and its huge black hole", by director Thomas LUCAS.

– Secrets of the Cosmos : The Higgs Field Theory).

– On the origins of the universe

– The Dark Ideas of Physics

– The keys to the universe 4 - The origin of water on earth

– "The mysteries of the Universe beyond the cold", written by Colin CAMPBELL

– "Discourse on the origin of the universe", Etienne-Klein – 2010)

– AurélienBarrau (Cfr : conference : What was there before the Big-bang ?".

Behind the cover

This book is not a religion textbook, but rather the results of 10 years of research that led me to discover that the biblical creation story is true. Let's not forget that this story comes to us from the papyrus of the 15th century B.C., whose author attributed all its manifestations to the divine will. By putting the name of God in evidence or out of the way, and scientifically restoring each verse, I found myself in a scientific scheme similar to the Model of String, developed by the Italian physicist Gabriele Veneziano.

In my research, I discovered that all the great structures that scientists have discovered and continue to discover in the universe were already revealed in the biblical creation scheme.

This scientific reconstruction of the biblical scheme of creation is the subject of a new cosmological model that I have named "The Universal Cosmological Model".

The first five chapters prepare scientists to understand the biblical language and non-scientists to understand the language of cosmology. This leads us from chapter VI to the discovery of the structure of the pre-Big Bang to the Big Bang, as well as to the scientific restitution of the antique cosmological models.

I leave it to those in the field to judge for themselves the results of my research, which is spread over the other volumes of the book series "Verses of the Scholars".

Sommaire

Librinova
www.librinova.com